NOUVEAUMANIA:

The
American Passion for
Novelty
and How It Led Us
Astray

ALSO BY TRUMAN E. MOORE:

The Slaves We Rent
The Traveling Man

NOUVEAUMANIA:

The
American Passion for
Novelty
and How It Led Us
Astray

TRUMAN E. MOORE

Random House • New York

Copyright © 1975 by Truman E. Moore

All rights reserved under International and Pan-American Copyright Conventions. Published in the United States by Random House, Inc., New York, and simultaneously in Canada by Random House of Canada Limited, Toronto.

Library of Congress Cataloging in Publication Data

Moore, Truman E.
Nouveaumania: the American passion for novelty and how it led us astray.

Includes bibliographical references.

2.   Technological innovations—Social aspects.
3.   Consumers—United States.   I.   Title.
E169.12.M66   973.92   75-10278
ISBN 0-394-48363-4

Manufactured in the United States of America

2 4 6 8 9 7 5 3

First Edition

TO PEG

. . . the security of the familiar—the terror of the new; the unhappiness with what is—a despair that it might be better.

From the Act of Penitence,
by Reverend John B. Macnab
The First Presbyterian Church

New York City
October 6, 1974

Contents

NOUVEAUMANIA:

The
American Passion for
Novelty
and How It Led Us
Astray

Introduction

Liv Ullmann said American films glorified the fake, the phony, and the perishable and that in America people did not know who they really were. Such a statement may be easily dismissed as a movie star's casual remark unsupported by documentation, but it contains a recognizable description of American life not easily applied, say, to the Germans, the English, or the French. Why do such unflattering characterizations stick to us?

It is very difficult to generalize about a people with any degree of sensitivity, and impossible to do so with absolute accuracy. *Nouveaumania* is an attempt to examine American character within the confines of the difficult and the impossible. We are a complex, creative, powerful nation. At the same time we are often simplistic in our approach to life, and a little insecure. We are a puzzling mixture of conflicting and contradictory traits.

*3

The picture of Americans drawn here is based on my own experiences and impressions gained from being one. I have also used the observations of others made available in conversation, books, articles in professional journals, popular magazines, and newspaper reports. The validity of my account, like that of Ms. Ullmann's characterization, rests on recognizability, on the reader's discovery of some part of his own life in what I have described as being a part of mine.

I am indebted to many people who aided me, often unknowingly, in the preparation of this book. Too often at social gatherings my conversation turned to interview and if I bored or offended friends, neighbors, and acquaintances, I publically apologize.

I would also like to thank Professors Richard Sennett and George Novobatzky, who read the manuscript and made many helpful suggestions but who stand innocently clear of my conclusions. My wife, unlike the wives of so many writers, typed very little of the manuscript but was invaluable as a source of ideas, information, and encouragement. My other editor, Robert D. Loomis, was unfailingly generous with his time and attention without which this book would never have been formed.

T.E.M.

1

The Postwar Promise

When I was twelve, my father and I had a running argument about the possibility of sending a rocket ship to the moon. Impossible, he said, because of the tremendous pull of the earth's gravity and the weight of the fuel a rocket would have to carry. To me, these were mere details. I argued on faith.

Like many Americans, I believed that new inventions and discoveries would solve everything, not only in getting to the moon but in bringing perfection to earth, an idea which, like all dangerous notions, was at least half true.

I felt lucky to have been born when I was. I had hit the last of the Depression, seen the end of the war, and since Our Side possessed a clear superiority in moral character and technical genius, I anticipated experiencing the application of American goodness and inventiveness to peacetime.

When the war ended in 1945, I was ten years old. I had

grown up on war news, victory gardens, scrap drives, and patriotism. I had never seen, to my memory, a new car, or with few exceptions, a new anything. The really new things were all in the military, like radar, the Norden bombsight, jeeps, and finally, the Bomb.

I believed, as did my friends, that the further into the future one conjectured, the better things would be. I heard a lot of talk about "someday we'll have . . ." in which the wondrous marvels of the future would be elaborated. The realities of that future were already apparent in the implications of the "Adam bomb," as I first misunderstood it to be called, but they had not dawned on me or most adults.

I read somewhere that there was enough energy in a glass of water to power a steamship across the ocean, and my vision of the future accommodated itself to this expectation. It was not until years after the war that I had nightmares about atomic war, nightmares the next generation would also know, but without the tempering thoughts of the miracle in the glass of water.

"Synthetic" became a big word, more easily grasped than the "mystery" of the atom that editorial writers were so fond of discussing. It was expected that as natural resources became exhausted, synthetics would be ready to take over, providing an artificial cornucopia to pour forth abundant substitutes for any shortages in the supply of rubber, minerals, oil, gas, wood, and metal, a notion that made concern for conservation a fetish of those opposed to progress and blind to the promise of tomorrow.

The world of plastics was beginning to be revealed to the nation's wondering eyes, and each new miracle was awaited with eager innocence. I read an article in *Life* magazine about a tube of plastic goo you could blow bubbles with that didn't burst. I could hardly wait until the dime store in my small town finally got around to stocking the stuff, only to discover that it had a noxious odor, that the bubbles were misshapen and useless because they did not float in the breeze, and that they had none of the redeeming iridescence of soap bubbles dancing through the air like butterflies.

Television was the biggest postwar miracle for most people; though it may only have been a grainy, snowy, rolling image of Milton Berle, it was enough of a spectacle to keep millions staring at bluish confusion, hoping to catch a fleeting image of a human form. I saw television for the first time in a motel near Washington, D.C., on a vacation trip with my sister and her husband. The show was a juggling act, but the picture was not sharp enough for me to tell what the man was juggling. At home, reception was so bad that few people bothered to own a set, and by the time the inevitable antenna sprouted on our roof, I was a young man off to college, with no time for watching television.

I lived—from the middle of the third grade through high school—in Myrtle Beach, a resort on the South Carolina coast that drew vacationers from the distant cities of Charlotte and Atlanta. But remote as we were from the larger world, its miracles came to us sooner or later. Our future would share the postwar promise. It would, for example, be free of flies and mosquitoes because DDT had been made available for civilian use. I can still hear the muffled puttputt of the compressor borne by a brown pickup truck creeping through the streets at sundown and covering us with a pesticide fog. I liked the smell of it. It was fun to breathe it and run through it and know it was deadly to other things, but that I, like Superman, was immune.

When I was growing up, the streets were mostly powdered dirt and sand, except for Ocean Boulevard, the Kings Highway that ran parallel to it, and the business streets in the center of town. The unpaved streets in the neighborhood where I lived were cool to walk barefoot on in the summer and hard enough for football and baseball during the fall and winter, when there were few cars. They were ideal streets for walking to the beach, and they imparted a friendly and relaxed atmosphere for the thousands of inland people who came each season, most of whom stayed in the weathered guesthouses and cottages rented for the month or the summer. There were motor courts out on the highway for weekenders and Florida-bound Yankees, but the tourist accommodations near the ocean were more homelike, less flashy.

The streets were finally paved, I believe, without any thought given to the possibility that unpaved streets were better suited to the needs of a beach community, particularly when the paving was a ribbon of asphalt laid down without sidewalks.

The first time I went back home after being away for several years, during which the town had experienced an enormous spurt in growth, it seemed to me that those paved streets had let in the highway ambiance. The summer cottages that lined the streets down to the ocean had begun to be replaced by motels adorned with all the gaudy, flashing signs they could support. Starved for parking places, the motel men had paved every inch of available land, bulldozing the sand dunes, running their parking lots from the edge of the ocean to the edge of the streets, making a solid cover of black asphalt that turned liquid and sticky in the summer sun, hot beyond belief. Where once you could walk to the beach in comfort, you now had to walk in a tarred no man's land.

The town I remembered had covered itself with asphalt, plastic, neon, and the postwar spirit. A friend of mine who wrote for the newspaper said that the whirl of air conditioners in the motels had finally "drowned out the sound of the breaking surf."[1]

One of the oldest and largest stores, Chapins, was once the center of Myrtle Beach. The style of the store was typical of low-country architecture, with archways at the entrances, and a stucco and tile exterior. The store covered an entire block and was lined with large palmettos that shaded the sidewalk with blue coolness in the summer months. As the years passed, the center of town began to lose business to the shopping plazas springing up along Kings Highway. In an effort to keep up with the times, Chapins remodeled the store, replacing the harmonious arches with a plain metal front, severely modern and square. The palmettos were cut down because they were blocking the sidewalk. The heart of town disappeared and its place was taken by another shopping center.

The other landmark to disappear was the Ocean Forest Hotel, built in 1927 in the manner of the grand ocean-front palaces of other East Coast resorts. It was recently dynamited into rubble to make room for a more modern structure. There was some regret at its passing, but most of the newspaper coverage dealt with the techniques of demolition.

Perhaps the most striking imposition of the new times upon the old beach atmosphere was the fast-food franchise mania. The drive-ins I remembered were strictly local efforts that usually consisted of a modest building in the middle of gravel lot where young people could swarm in the evening for impromptu socializing and a few surreptitious beers, if they were underage.

On my return to the Beach I was surprised to see the wild assortment of food outlets that seemed to have been shipped in fully constructed, with yellow plastic mansard-modern roofs, wraparound glass walls, and boasting names and ersatz decorations taken from the American West, the South Seas, the plantation kitchen, and triumphs of mass production ("9 Billion Served").

The whole country, like my hometown, has covered itself with cultural imitations shouted in neon, stuffed itself on fast food, and developed a madness for tearing down, paving over, bulldozing, dispersing, growing.

We are now living in the future we looked forward to at the end of the war. Many of the expectations of those days have come to pass. Nuclear power is replacing fossil fuels, synthetic materials are everywhere, television is in focus, our reach exceeds the moon. The miracles that Americans dreamed of twenty-five years ago have come true and many of us are not pleased with them. We read the announcements of new wonders with apprehension, and beginning most significantly with the SST, we sometimes reject them, or like DDT, we ban them.

The nostalgia for the forties and even the thirties evident today is due in part to a desire to return to a time of our lives when the promise had not been fully realized, when our dreams had not turned into nightmares, before the term "postwar" came to mean something slapped together, such as cars, houses, and appliances.

In the rush to supply the country with all the consumer products it had been without, to provide the housing needed by returning veterans and their families, to cope generally with the changeover to a peacetime America, the demands of the moment made expediency more important than quality. Levittown was rising from the potato fields of Long Island, and though "it looked like houses in a Monopoly game" from the air, young families, bearers of the postwar baby boom, rushed in, glad to have a place to live, even though there were "no trees, no schools, no churches, no stores, no private telephones," and when it rained, everything turned to mud.[2]

The postwar spirit exaggerated some of the best and worst national characteristics. The veterans who came home unimpaired threw themselves into civilian life with all the gusto of the pioneer, the traveling salesman, and other intent and energetic mythic Americans. The GIs remembered the Depression I had heard about or only dimly recalled, and had witnessed the war I had only read about. Having experienced a full measure of poverty and war, they wanted security and peace. It is the result of these urgent needs that has shaped the country, and remembering this, it is possible to look at America critically, but with sympathy.

The shape of my town and the country was molded by a postwar spirit that was individualistic, expedient, concerned with measurable results—not style and manner—and above all, it was innovative. It was the early American pioneer spirit, bent out of shape by time, wars, depressions, and rampant growth, demanding more than ever that new technology give fast answers.

I'm not home any more to argue with my father about things like going to the moon, but when I was visiting one summer I asked him if he felt we had fulfilled our promise as a nation since the war, and if he was satisfied with what we made of ourselves. He said he was optimistic but disappointed. We were in decline as a nation, and the freedom of the individual was in serious jeopardy. It was sad to hear him say it, but then, I don't argue on faith any more either.

Plymouth Rock,
oldest symbol of America's cultural transplant,
fenced off on a narrow beach between
the bay and the highway.

2

The
Cultural
Transplant

The malaise running through American life is tied to the breaking of the postwar promise, but its root system goes farther back in time, to the beginning of an even greater promise when European culture was transplanted to American soil. It was during this process that the colonial immigrant to America stepped out of the historical framework that had shaped the thinking of Western man for centuries.

The building of the American character began when the first settlers, confronting the wilderness, discovered that their ideas and fund of practical knowledge did not usefully apply to the conditions of the New World. Europeans have often looked at us with great insight on this point. Henry Bamford Parkes, an English historian, wrote: "The American acquired new attitudes and learned to see the world in a new way. And the nationality he created became a vast experiment in new social principles and new modes of living . . .

By crossing the Atlantic, the American had asserted a demand to be himself; he had repudiated the disciplines of the class hierarchy, of long-established tradition, and of authoritarian religion."[1]

Having freed himself from the bondage of European traditionalism, the new American bound himself to an illusion of his own making, to which I have given the name nouveaumania. Its origins lie at the heart of American tradition, beginning with a new country in a new world, a new society, a new type of government, and a national beginning at a time when rapidly changing technology was becoming a new fact of life.

Nouveaumania developed from the belief that novelty was a panacea. Faced with the need to innovate or adopt new ideas to survive, the American settler made the search for new ways of doing things a part of the American tradition. In recent years we have seen the results of the distortion of this tradition in the adulation of new technology, a disinterest in the past, and a pervasive sense of impermanence.

The first Americans were, in a way, social mutants who grew differently from others in the culture that spawned them. They thrived in an atmosphere which older, more established societies found poisonous.

Change was our heritage, the turbulence it created was our norm, and the New Way our imperative. We put our faith in it, whatever it was, and on the inventions, ideas, and machines that opened new era upon new era as America grew with the Industrial Revolution. We lacked, during our growth as a nation, a period of tranquillity in which to absorb the shock of transplant from the older, more ordered societies of Europe to the wide open spaces of the New World. Our formative development as a nation took place in a state of social flux we came to regard as normal.

One of the great scenes in American consciousness is that of the pilgrims standing on the cold New England shore ready to impose their notions of a new and perfect society upon the blank forest.

Throughout American history, restless and discontented men have sought new opportunity, new beginnings. Historians have seen this impulse as a manifestation of our courage and love of adventure; they have also called it escapism and a lack of commitment, a tendency to move along rather than to get along. Either way, it convincingly bespoke the willingness of the American, and the immigrants who continued to come here, to risk the new and the unknown.

What has happened in America over the years is that we have gone from a readiness to live with necessary change to a hunger for change as an experience or an entertainment. We have become addicted to novelty, infected with nouveaumania. It was in our mother's milk. Our forefathers were stimulated by it in those crucial days of national beginning and now we must have it—not just change for the better, but for itself, because we have come to think that life must change constantly. Tranquillity and balance are the same to us as stagnation, and both mean the death of the soul.

The attraction to the new means a lack of interest in the old—in old ways of doing things, old buildings, old people, old ideas, old traditions. The American impulse is to grab the latest thing, to seize the new and heave the old, an outgrowth of early survival traits. Some societies are characterized by complacency and traditionalism; American society by its turbulence. What is not new and better, or instantly improvable, is deemed disposable. The reform spirit is part of our character and has made America, in the words of Woodrow Wilson, "the only idealistic nation in the world." Idealists are not easy to live with; they can make that sense of self-esteem so vital to the well-being of self and nation hard to capture.

We have always regarded as fools those who did not see how fast the world was changing, like the man in early America who, having amassed a fortune and wishing to protect his progeny from their own folly, stipulated in his will that none of the stock in which he had invested should ever be sold. The dividends would be sufficient to provide them with comfortable incomes, for he had invested wisely in companies that made basic products for which there would always be a demand—wagons, candles, horse collars, and the like.

In America, we fear obsolescence not only in commerce and industry, but in people and in the nation itself. The English belief in the continuity of life is reflected in the sayings, "There will always be an England" or "England will muddle through." Americans have an overdeveloped sense of perishability. In a single week, three stories appeared in the *New York Times* about the impending economic collapse of Italy. No one seemed to get excited about it except the Italian finance minister and us.

Our national sentiment, expressed by the Founding Fathers, the Gettysburg Address, and the usual newspaper editorial, is a sense of national impermanence. It is said that a woman stopped Benjamin Franklin as he was leaving the Constitutional Convention at Independence Hall and asked him, "Well, Doctor, what have we got, a republic or a monarchy?"

"A republic," Franklin replied, "if you can keep it."[2]

Keeping it in 1787 may have been a matter of real concern, but after almost 190 years we continue to see our national existence as problematic, even in the most tranquil of times. We are constantly obsessed with issues and events that are deemed capable of destroying the country as if it were on the same fragile footing as in Franklin's day. This sense of national impermanence arises from the knowledge that our own taste for change robs tradition of its place in creating a sense of order upon which national security, as well as personal security, ultimately rests.

Old World traditions were once at least examined for salvage value. There is a famous passage by De Tocqueville about what he called the philosophical method of the Americans: "To evade the bondage of system and habit, of family maxims, class opinions, and, in some degree, of national prejudices; to accept tradition only as a means of information, and existing facts only as a lesson to be used in doing otherwise, and doing better; to see the reason of things for oneself, and in oneself alone; to tend to results without being bound to means, and to strike through the form to the substance . . ."[3]

This is, ideally, how we like to think we still behave, forever struggling between ocean and wilderness, shaping a new world by our spontaneous, fresh action.

In industrial America, old traditions are junked like worn-out cars and appliances. The rate of change has accelerated to the point where today people feel that nothing seems to work as well as it should, that they have somehow lost control over their lives. What is missing is the sense that the present grew logically from the past and that tomorrow will be continued from today.

Almost daily, hourly, one hears or reads about the widespread discontent in American life. The causes are variously attributed to poverty, affluence, inflation, racial tensions, war or the threat of war, and pollution. Our discontent has been given many explanations, but not since Hubert Humphrey ran for President in 1968 has anyone tried to tell us how happy we were. His joyous campaign theme was one of the biggest political miscalculations of all time.

When foreign visitors wish to pay us a compliment, they nearly always refer to nouveaumania in a positive manner. When New Zealand's ambassador of ten years went home, he was asked for his impressions of America. "The main thing I go back with now is a sense of the enormous creativity, adaptability and energy of Americans. At times I wondered if the United States was a country on the decline, but now, on balance, I think it is a country that is bursting into something new, rather noisily and argumentatively, but something new."[4] Being a diplomat, he did not say what it was. A recent visitor from China spoke of our "constant and intense search for new ways of doing things."[5]

This tradition of inventiveness, bolstered by optimism, has indeed served us well, but it stands today in danger of deteriorating into a penchant for gimmickry. We are losing the distinction between the courage to innovate and an infatuation with gadgets, between real improvement and a frantic search for quick panaceas.

In America today, the readiness to accept real change, which is always essential to healthy growth, has been corrupted by the superficial or merely novel and has contributed to the feeling that as a nation, we don't seem to be dealing with our problems. We need better education for our children, jobs, a clean environment, safe transportation, a sense of personal safety and well-being. Instead we get more weapon systems, a plethora of junk products, superhighways, a general sense of decline, and photographs of Mars.

America, we believe, is the land of the new, the home of tomorrow. One of the most distasteful claims of Communism is that it is the wave of the future. And no Russian impertinence was so unforgivable as that first Sputnik. Although many Americans later claimed indifference to the successes of our space program, nothing would have been more offensive to our pride than to have seen the first moonrise with some other human than an American gazing back from that cold place.

Our nouveaumania is not limited to a taste for gimmicks but seems to cling to life, expressing itself in symptoms of the failure of reason, largely in our general discontent and minutely in such things as the plastic ersatz with which we surround ourselves. A reviewer who was impressed by the authenticity of the set in a television play observed: "There is a dish of artificial fruit, a vase of plastic flowers . . . It is an honest room where millions of Americans live . . ."[6]

We came to live in that honest room as a result of a hurried-up history which created our new nation with a society spreading across the continent like a shallow-rooted vine, covering everything, grabbing anything. As a young country we had an impatience with the past and a predisposition toward change, novelty, and improvisation nurtured in colonial times, when contact with the stable, civilized world was physically and philosophically cut off, and survival depended on the ability to adapt and search for new ways.

From the beginning Americans felt that they were different from other people—freer, less encumbered by the bonds of age-old habit and tradition, possessed of a new and special character. But the excesses of that character have affected our perception, our ability to discern the valuable from the worthless, and have left us with a junk-cluttered society we are told can survive only if we produce increasing amounts of increasingly worthless products. It has, in effect, created the artificial American, a creature who inhabits the psyche of every one of us and whose unchecked behavior has led to what must be recognized as a form of insanity, that of nouveaumania, the frustrated colonial-frontier spirit that never rests but continues to haunt the land like a cultural poltergeist—banging traditions, rattling customs, screaming for tomorrow.

3

Buying
Happiness

It has often been asked how it happened that America, founded on such idealistic principles, came to be so materialistic. How did the pursuit of happiness become confused with the purchase of happiness?

There was, from the beginning, a submerged streak of materialism running through American life which was combined with religious fervor to produce what Max Weber called the spirit of capitalism. Prior to the time that the industrial cornucopia ran wild, the energies of manufacturers were directed at meeting the demands for goods by a growing population rather than at creating needs for that population.

The ascendancy of productive abilities of factories over the material needs of the public led businessmen to put strong pressures on individual citizens to become what we know today as consumers. The techniques of high-pressure salesmanship had been worked out and put into practice as early as the eighteen-nineties by such men as John Patterson of the National Cash Register Company, but it was not until business fell off after World War I that these techniques were seriously applied by almost every business of any size. There was widespread concern, particularly during the depression of 1920–21 that general overproduction would remain as a permanent condition even after the business crisis had passed. So dawned the Marketing Age, or the Age of the Salesman, with its sales quotas, territories, conventions, contests, and general hoopla.

Frank Sullivan, writing in *Our Times*, his six-volume account of the first quarter of this century, noted a "mood of irritation" among many Americans and a feeling that they were being pushed by forces they could neither see nor identify, feeling vaguely that their freedom of action was being crowded by "some invisible power." That irritating power was generated by the rising sales pressure building up all over the country in response to the growing need of business to dispose of excess production.

The chief proponent of unyielding mass production had been Henry Ford, but when the last Model T rolled off the assembly line, it was widely recognized that mass production did not in itself produce mass sales. In their urgent need to find buyers for the growing quantity of goods that the factories were producing, business managers naturally, and perhaps unconsciously, fell upon the theme of novelty. In the future their products would always be new products. However much this might have suited American nouveaumania, it ran counter to the basic concept of mass production. A constant parade of new products was inconsistent with the economies of large-scale manufacturing, but ways were found to make gradual changes in the model line, more cosmetic than substantial.

Paul Mazur, a young New York banker, reacted to this in 1928: "Bathtubs appear in stylish shades. Dishpans no longer are plain Cinderellas of the fire and ashes; following a wave of the hand by the fairy godmother—style—they now appear resplendent in blues and pinks. It is the vogue to have tinted linen—this year green and peach; ncxt year, the sales manager hopes, beige and blue. . . . Automobiles change with the calendar. Last year's offerings are made social pariahs, only this year's model is desirable until it, in turn, is made out of fashion by next year's styles. Furniture, clothing, radios, phonographs, tumble from the fertile minds of scientists and designers in floods that drown the sales possibilities of products that already stagger with infirmity at the age of one year."[1]

By the constant suggestion that to buy the lastest styles was to be up to date, producers harnessed the desire of the American public to participate in new happenings, to keep up with the times. As American salesmanship and advertising warmed to the task of forming this vital link, the public's eye was eventually trained to watch for the latest manufactured wonder.

We came to regard new products as the prime indicator of progress, and the acquiring of new products as a way of personally making progress. This was in keeping with the older American belief that all change constituted an instance of progress, which was indisputedly good and therefore desirable.

While the Depression and the war put something of a damper on the proliferation of new products, intimations of future goods and services filtered through the media, mainly in magazine advertising and articles. In 1945, before the war ended, the pages of the big magazines furnished the material for our daydreams of the future. Wartime research had created the potential for thousands of new products and services. They promised a world beyond tomorrow that would be wonderful—beyond our wildest dreams, as they used to say. In looking through the magazines published during the last months of the war, you can sense the anticipation, the eagerness to live the dreams depicted in those colorful pages.

A 1945 Goodyear ad in *Life* depicted the sky filled with

small helicopters and a man standing on top of a building hailing a sky-cab. The helicopter was widely believed to be the family car of the future, and the idea that it might be the taxi of the future put no strain on popular imagination. Another series of ads in *Life* showed the innovations the automobile driver could expect in the future, including such aides as "fog-melting" headlights and dashboard instruments that would inform the driver of traffic conditions ahead of him.

The dirigible was seen as the "flying hotel" of the future, the natural replacement of the ocean liner, making the crossing to Europe in forty "glorious" hours, a vision that missed the frenetic spirit of the Air Age by a wide mark. In one ad the Martin Aircraft Company presented an artist's conception of a cargo plane transformed into a spacious airliner, with lounges, bedrooms, and dining booths. Before the war the Boeing 314—the Pan Am Clipper—had briefly provided the transocean passenger with similar accommodations. It was natural to envision even grander postwar flying hotels, whether airplanes or dirigibles, in which families were shown in spacious cabins, perhaps seated around the table while the waiter served an eight-course meal and the sights of the world drifted past the window.

For the home, manufacturers assured the housewife that with the new appliances to come, "Most of the drudgery and routine of running a house becomes merely the flicking of a switch or the pressing of a button." Hotpoint and Kelvinator, totally occupied with war production, talked about the wonders of the postwar kitchen, RCA talked about television, and Philco about freezer-lockers, although adding that "This is not the time to speak of them . . . nor the time to suggest their delivery is close at hand." General Electric promised a new kind of sleeping comfort with their new electric blankets, which were based on the same technology that made high-altitude flying suits for pilots.

General Motors felt that America was ready for the "Journey into Tomorrow," in which "Men of science are moving forward with new methods and improved products . . . with the return of peace, ever-better ways of doing things will result in a flood of new benefits to mankind." It is a tribute to the commercial astuteness of this most American corporation that it captured the fantasy of buying happiness as it and the other consumer companies fanned the flames of nouveaumania.

Even the most overblown praise of the future seemed to have a factual base. News reports of fiber glass, epoxy glue, phosphors, the electron microscope, came in a steady stream as discovery and invention were made known to the public every day, not infrequently as new products. Firestone promised a better nights sleep, after the war, on a mattress made of a new material called latex foam, adding, with a combination of pride and regret, "Foamex is at war now, shielding men and instruments from concussion."

As the factories began to retool for the consumer battle, the public was eager for new things to buy that carried the excitement of the future, and for the first time in years, there was money to buy them with. The public's romance with novelty blossomed like young love. Sleeping on foam rubber, pushing a button to end household drudgery, hailing a passing helicopter—it all conjured up a vivid picture of the new and glamorous life.

We had, for all practical purposes, defined our postwar future in materialistic terms. We saw our destiny as shaped by things one could shop for. The end of the war left us no restrictions on how much happiness we could buy. New products, new ways of traveling, cooking, sleeping, eating, were seen as progress which was only good. In our rush toward tomorrow we ignored other possibilities for improving the quality of our lives. We entered the postwar era as ardent consumers, impatient for our long-delayed honeymoon with the future.

New!

An American editor returning recently from a twelve-year stint in Paris wrote that in Europe they say "It's good because it's old." In America, he noticed, we say "It's good because it's new."[1]

About the only items that are advertised here as unchanged for long periods of time are a few brands of liquor and beer. Everything else parades under the banner of "new," which is the strongest word in the language of advertising. I took a course in newspaper advertising in college which turned out to be almost devoid of theory, with the single exception of the professor's firm conviction that the word "new" should appear in every ad. It was sound advice. American industry expends great effort and imagination to find real or imagined excuses for using that magic word on every package.

Always curious to try a new product, we seem to feel per-

sonally enhanced by its possession. It is not a new model
car we buy, but a new model of ourselves, identifying, as
motivational psychologists tell us, with the image of the
thing we have purchased, whether it is a soft drink, a can
of soup, or an automobile. These new images we buy lift
our fallen hopes, offer us different versions and new models
and styles of ourselves, i.e., a "new you." They promise us
anything, even if they give us, finally, nothing.

Recently an Israeli student told me that one of the first
things she noticed in America was the number of products
that claimed to make her a "new person," a transformation
she found unappealing. Russell Baker, in his column in the
New York Times, caught the essence of this in his observa-
tion about a milk ad on television that promised "a new
you coming," which implicitly assumed that the viewer
"disliked himself so thoroughly that the prospect of a 'new
you' arriving after milk ingestion would be pleasurable to
contemplate. . . . Flawed though the present me may be,
my experience of that inescapable American marketing phenomenon, newness, compels me to believe that the new me
would almost certainly break down easier, wear out faster,
emit more ghastly pollution, and contain more plastic than
the present me."[2]

The marketing of new products calls for the greatest in-
genuity in those companies dealing in such basics as gasoline, soap, household cleaners, paper products, and similar
goods where the creation of a new feature is more dependent
on the imagination than on engineering, or in standardized
products such as soft drinks, crackers, and other food
products made according to unchanging recipes. It is in the
promotion of these products that the greatest amount of
fantasy is incorporated into the advertising, and it might be
said that the greater the smoke, the lesser the fire. The Pepsi
generation, in terms of pure symbolism, may survive as the
representation of meaninglessness in our time.

The artificiality of the new products race to market is
not only in the products, but in the race itself, sometimes

depicted literally on television as a duel between catsups to pour, towels to absorb, toilet paper to unroll, aspirin to dissolve, cars to run out of gas, and spots to disappear.

The nouveaumania that possesses most Americans will not only tolerate a good deal of make-believe in perceiving the new among the fanciful presentations of the old, but will readily accept substitutes offered in the name of progress, providing a suitable atmosphere for the introduction of texturized polyester pants, vinyl shoes, polyurethane furniture, acrylic bathing suits, Sani-Gard underwear, Durapuff mattress pads, Serofoam mattresses with Tufflex foundations, polystyrene, pearlescent toilet seats, electric fireplaces, plastic trees, artificial grass, fortified factory-made food, and a host of other products that look like, taste like, feel like, smell like, and sound like something other than what they are without the natural faults of chipping, staining, running, wrinkling, souring, flaking, molding, peeling, or costing as much as the real thing.

By habit, the American consumer approaching the shelf in the store looks for the word "new." Believing that technology is constantly at work making the world more wonderful, he assumes that the latest product will incorporate all recent and beneficial advances.

The number of products containing an important new feature is not large. Most are familiar items with something tacked on to appeal to the consumer's passion for novelty. Gasoline with ZX100, automobiles with tail fins, lime-scented shaving cream, or detergents with little blue crystals.

The consumer, though an easy mark for a new product, is quickly disillusioned and moves on. A Don Juan with a shopping cart, he leaves 80 percent of the new products he samples broken-hearted failures, destined to go the way of —remember?—Gablinger's beer, Red Kettle soups, Corfam, the aerosol toothpaste, and other Edsels of the consumable set, most of whom will die quietly in their Midwest test markets.

Brand loyalty has become only a nostalgic reminder of

bygone days. I can remember whole families who always voted Democratic, drank Coca-Cola, shopped at the A&P, and drove Fords, to whom off-brand was anathema and brand switching a display of weak character. The constant lure of novelty has destroyed constancy as a virtue in consumer morality.

American industry today spends $15 billion a year on new products, creating a frenzy of activity that cranks out 6,000 new items in the drug and grocery business alone. During the present decade 120,000 new soaps, foods, snacks and other products will be put on the market, and 100,000 of them will fail, even after passing a series of elaborate and expensive market studies and tests.[3] The consumer apparently has difficulty experiencing that many new "needs," yet the survival of almost every company offering goods and services to the public depends on new products—that is, on stimulating new consumer "demands."

The idea whose time had come at the end of World War II was that products were progress. General Electric used to say, in all their advertising, that progress was their most important product. The head of their research lab once put it a little differently when he said that engineers should produce obsolescence. Perhaps the accurate statement most companies could make is that products are their most important progress.

Today that idea is losing credence with the public. Companies that rush to market with another me-too product, or as one marketing executive put it, with a time-consuming, labor-saving device, are sustaining heavy losses in new-product failures. In spite of computer analysis, sophisticated marketing plans, and sampling techniques, most new products don't make it because they don't do anything for the consumer.

This was the prime failure of Corfam, the "leather" of the future, and one of the most spectacular marketing failures of modern times. In replacing leather with plastic, DuPont could point to a number of benefits. Hides are irregular in shape and size and need extensive handling during the tanning and manufacturing processes which transform them into shoes, belts, and handbags. From the consumer's point of view, several real benefits were pointed out. Corfam was waterproof, required little upkeep, and was supposed to be less expensive than leather, although producing a semi-permeable artificial leather proved more difficult and costly than expected.

Even when the production problems had been solved. DuPont encountered the insoluble problem that finally brought ruin. Shoe manufacturers and customers didn't like the way the stuff felt. And in spite of DuPont's insistence that their polymeric film was more porous than leather, customers complained that artificial leather felt hot.

I bought a pair of Corfam shoes—the clerk told me they were "better" and I believed him—and DuPont may have for free my analysis of what went wrong with their product. The shoes cost as much, or more, than leather shoes, and they wore out just as fast. They always looked as if they needed a good buffing, even if you had just wiped them off with a cloth, and they never took a shine properly with polish. While it is true that they were waterproof, it has never been my habit to walk in water with my shoes unprotected. And like all those other people, I never liked the way the stuff felt, either.

Exactly what it is about products that make them fail or succeed in establishing a place in the market is certainly unknown ahead of time, and official explanations for failure after the fact have a wonderland quality. General Foods, for example, saw its greatest innovation—Maxim's freeze-dried coffee—beaten in the market place by Nestlé's late-entry imitation, Taster's Choice. G.F. attributed this to poor marketing strategy. When the company entered the single-serving dessert market, it did poorly with its Jell-O Pudding Treats, but explained that it was "too late" to catch up with Hunt's Snack Pak, which had entered the market first. By this reasoning, Maxim should never have been overtaken.

The failure of G.F.'s Toast 'Em Pop-Ups pastry line was attributed to "aggressive competition," hardly an unusual circumstance, and the $39 million the company lost on its Burger Chef chain was attributed to a soft economy, although other fast-food operations did well enough during the same period.

While post-mortems of such disasters can make interesting reading, generalizations useful for future ventures are hard to extract. When new products offer the consumer no important benefits, their survival must depend on the split second when whim and impulse direct the buyer's hand. No one has yet found any reliable guide to that fateful moment.

In 1971 General Foods reported a loss of $83 million, which they managed to refigure and announce as a loss of $46.8 million. The bad year cost the president his job, produced several papers at the Harvard School of Business Administration, and caused Wall Street observers to ask: "Where are the new products? Where are the new ideas?"[4]

The exploration of the consumer's nouveaumania is not the only force that generates new products. Robert Townsend has pointed out that the chief executive, in his ongoing need to protect his flank, causes the clamor by titillating the board of directors with "half-baked plans and untested products" that "are accelerated to the boardroom and served up predigested and oversimplified."[5] The premature result is another product added to the pile of shoddy goods currently sowing the seeds of a really serious consumer revolt.

There is, among appliance makers in particular, a Buck Rogers bias in which the "products are progress" fantasy has lost its last tenuous grasp of reality. Not too long ago, for example, *Business Week* magazine asked a group of industrial designers to fashion the can opener of the future. The most striking prototype was a real beauty: it stored the unopened cans in such a way that the housewife could dial her selection and push a button, whereupon the selected can would automatically be opened with a laser beam, the contents emptied into the waiting saucepan, and the empty can pressed flat and stored for recycling.

The push-button paradise envisioned for the kitchen of tomorrow raises some questions. When the laser malfunctions and refuses to open a can, or worse, opens a hole in the counter that is basement-deep, what happens until the repairman-technician arrives? Perhaps there is a small drawer containing a hand-operated opener to be used in emergencies.

How much are people of the future, still eating out of tin cans, expected to pay just to perform this mundane task? How big will the kitchen be if the can opener alone is three feet tall? Will a housewife or househusband of the future, running a house filled with such gadgets, have to be a production manager and a mechanic?

It is becoming increasingly apparent to consumers that all machines break down, even machines not designed to break down. If we elect, through our purchasing dollars, to encourage continued emphasis on products that are increasingly complex and decreasingly useful, we are choosing a future that will not be the postwar dream come true but simply frustrating, a future conceived in terms of convenience but offering no real improvement in the quality of life.

The country has changed since we first were taught that we could buy happiness, that a man who was abreast of the world drove this car or smoked that cigarette. It is possible that we are beginning to tire of the new-product game, a thought that has also occurred to Ted Angelus, a marketing consultant and new-product specialist. "Are consumers no longer gratified," he wonders, "by a fresher breath, a richer coffee, a whiter wash, but searching for different values?"[6]

It is quite possible, likely even, and the search for those other values may lead us away from our traditional ardent support of the novelty-based market altogether.

5

It's
Sweeping
the Country

The American proclivity for fads is one of the more dramatic manifestations of nouveaumania. The fad is instant conformity. It sweeps the nation by the force of our passion for the latest thing and vanishes before our contempt for the passé.

When Clark Gable took off his shirt in *It Happened One Night* and revealed a bare chest, men all over America stopped wearing undershirts. It was a bad year for the underwear industry. In more recent times, Johnny Carson appeared on *The Tonight Show* wearing a turtleneck sweater, and later a Nehru jacket, and sales for those garments jumped off the charts.[1]

The fad is the march of progress turned into a blind rush, revealing that in the public's nouveaumania lies the secret urge to be like everybody else.

In its heyday *Life* magazine reported and, in effect, helped create fads and crazes that swept the nation, leaving thousands, sometimes millions, of Americans dancing in marathons, wearing zoot suits, swallowing goldfish, cramming into phone booths, and swinging a hula hoop. *Life*, as part of its war coverage, had spread the popularity of the Eisenhower jacket, the Montgomery beret, and the Furlough Nightgown. Amid the celebrations of peace, they noted that the first postwar fad was blowing bubbles. In five weeks, 50,000 jars of a glycerin-based liquid, used with a metal loop, were sold in Atlanta, where the fad started.

When *Life* noticed that in some parts of the country teenage girls were wearing white socks, girls in all parts of the country picked up the idea from the magazine and a new human category was created—the bobbysoxer. Girls saw more to the look than just socks. They interpreted it as a part of a larger style, of what the American girl had become in places closer to the center of American existence.

Television eventually took over from the big national magazines as the showcase of our nouveaumania. To the delight of manufacturers and the frequent despair of parents, it proved to be a powerful popularizer of fads, particularly among small children. Big moneymaking fads were promoted on the box and spread with unequaled frequency. In 1955 the Davy Crockett fad, lasting all of eight months, spawned five hundred items, which sold for $100 million. In 1958, in five months, $45 million worth of Hula-Hoops crossed the counters, and the Wham-O Manufacturing Co. reported profits of $4 million on sales of $10 million. During 1955–1959 the kids bought, or caused their parents to buy, $283 million worth of cowboy paraphernalia, most of it hawked on television.

The adult fads during the fifties became known as status symbols. Favorites of the time included the bone-dry martini, tranquilizer pills, pedestal furniture, attaché cases, mobiles, and the "think book" of the moment. The counterculture, equally faddish, was (as they said) like into Zen and "beat" poetry, beards, and dirty feet.

Bucket seats and car coats were big items for the million sports-car owners, and for the motorcycle crowd it was leather jackets for men and the crinoline skirt and saddle oxfords for women.

Novelty gifts draw on the faddish impulse to create a strange line of products. Even during the war years, the Chicago Merchandise Mart was laden with wax steaks, lobster toothpick holders, artificial fruit, and lawn frogs. One popular gift item in 1945 was a Mexican pottery casserole of Welsh rarebit drawn by a pair of porcelain horses, all perched atop a pink box of English biscuits. Gift catalogues, now a familiar sight in the morning mail, carry on the tradition with such modern inventions as canary diapers, inflatable playgirls, mink-covered toilet seats, and chrome-plated marmalade guards for toast. Victor Papanek, who gave these last examples in his book *Design for the Real World*, noted to his horror that canary diapers were selling at the rate of 20,000 a month.

There is a constant attempt to create fad products based on events—commemorative mementos of the moment—such as T-shirts, plates and glasses honoring presidential visits, space ventures, and national holidays. When Bobby Fischer won the world championship, one company produced a necktie with a chessboard on it in such haste that the pieces were set up on the board the wrong way. Following the publication of a best-selling book, there was a rash of seagull jewelry—broaches, pins, earrings, tie bars, barrettes, and rings.

The Bicentennial (or Buy-Centennial, as some called it) brought forth a huge number of commemorative fad products, the worst and most expensive being a Boehm porcelain plate bearing the image of a rather squat baby eagle, entitled "Young America, 1776," priced at $175.

The Watergate investigation produced a predictable flood of fad products such as the Watergate Watch (guaranteed against break-in), a Nixon watch with a picture of the former President on the dial, his eyes shifting back and forth sixty times a minute. There were T-shirts with slogans such as "Don't Bug Me," and a Watergate game in which cheating was permitted.

The most pretentious of the commemorative commercializations are the medals and plates issued for collectors and investors, a venerable practice that has recently become a fad industry. The occasion upon which a medal is struck might be a contemporary event, such as the end of the Vietnam war, or it might be for no apparent reason, such as a recent series depicting the battles of World War II and another on the history of the automobile.

The terms "heirloom quality" and "collector's delight" are frequently used to describe the medals and related ersatz objects of art such as a sculpted version of the Last Supper, souvenir spoons, cups and saucers, coin sets, and various items of machine handicraft, many of which are dated or numbered to add the illusion of scarcity and value. The Last Supper was advertised as having been sculpted— and I am sure they meant to say "cast"—in a material called "genuine Italian oxolyte" which, according to two geologists I spoke to, is not a natural substance and was not known to them.

Some of the commemorative medals and plates have indeed increased in value since they were issued, but the majority are worth considerably less than their original selling price. The Franklin Mint's first Christmas plate, issued in 1970 at the beginning of the craze, has increased in price from $125 to $450. But the market for similar items has been flooded since the fad started. According to *Kiplinger's* magazine, dealers advertised previously issued souvenir plates at higher prices to make it appear that the price had soared.[2] The Lincoln Mint's "Collie with Pups," which was first sold for $125, was advertised by dealers for as much as $200 but later was widely available for $35 to $40.

The rationale that shimmers beneath the surface of some of the advertisements is intriguing enough to examine in detail. For example, the Franklin Mint—the undisputed leader in the production of commemorative souvenirs—produces endless numbers of silver medals and plates in "limited" editions, the limit being established by a cutoff date for placing an order rather than by a predetermined quantity. One of their recent editions consisted of fifty silver medals depicting the paintings of Rembrandt—a curious choice, since his work is known for its light and shading, qualities that were completely lost in the highly polished metal. "How fitting," said the ad copy, "that the greatest works of this towering genius will now be captured in a magnificent series of . . . medals that will endure long after the centuries-old paint and canvas have cracked."[3]

None of the reasons given in the advertisement can be said to be rational inducements to buy. The series was totally artificial in its stated premises. As representations of Rembrandt's paintings, they were without merit. There was no aesthetic relationship between an oil painting and a bas-relief image of that painting on a two-inch disc of silver, albeit sterling. There was no occasion or reason for striking such a collection of medals. The assertion made in the advertisement that the medals were of "intrinsic" value because they were made of "solid sterling silver" is misleading. In spite of a 100 percent rise in silver prices since the medals were issued, the value of the silver is still worth far less than the medals sold for. As an investment, it would have been wiser to buy plain bars of silver.

And finally, as for the superiority of metal over paint and canvas, the Metropolitan Museum of Art assured me that an oil painting, if properly maintained, will last forever. It does not anticipate any deterioration in its Rembrandts any time in the next several centuries, if ever.

A strong promise of permanence underlies the sale of this kind of keepsake kitsch. The wording of the ad was couched in terms of the medals' enduring—yea, increasing—value, an appeal that seemed to play not only on fears of inflation but on the sense of rootlessness that pervades our novelty-bound society. Are there really people who would buy this set of medals so that they might insure themselves of access to Rembrandt's images after the decay of the paintings? Or do they desire to leave them for the enjoyment of their descendants after the processes of decay have destroyed not only the original paintings but also the millions of reproductions, any one of which would be superior to the small medals for the study or enjoyment of Rembrandt? Or is it supposed that the prospective purchasers look forward to the day when fire will sweep the world and he or his progeny will sit in a cave, sole possessor—along with the other limited number of subscribers—of the remains of man's cultural heritage?

Whatever fantasies the producers of commemorative fad products aim to tap must be built on the sense of a transitory, impermanent world and a deep-seated fear of its sudden demise, a fear addressed directly in a *New Yorker* advertisement for Swiss wood carvings: "Now you can collect tradition."

If keepsake kitsch has a value, it is not in any of its explicit features but exists entirely in the buyer's mind as the prospect of possessing something that will "endure," and by some magical process lend stability and permanence to the owner. The Rembrandt medals are themselves symbols, not the bas-relief images of whatever the Mint decides to merchandise. They, and the "heirloom quality" gift items and other fad products, are symbolic of our desire for verities, for something to hold on to and depend on in a world that changes too much.

Fads and crazes, while often harmless enough in themselves, are the chills and fever of our nouveaumania. These sudden seizures, while not unknown among the people of other nations, are more frequent, more extravagant, and more broadly based in America. They do their greatest damage not by the millions of dollars wasted on them, but by perpetuating our delusions.

6

The
Compulsion
to Imitate

Mrs. Potiphar, that social-climbing *nouveau riche* of the eighteen-fifties, was the fictional creation of George William Curtis in his satire *The Potiphar Papers*. She embodied the eagerness with which Americans imitated things they did not fully understand. Though Mrs. Potiphar was a little hazy on the details of Europe—she chattered about "Dresden, Vienna, and the other Italian towns"—she managed to follow the habits and styles she most admired. If she did so rather vaguely, few knew or cared.

When the Potiphars argued about building a new house, Mrs. Potiphar won by insisting that it was an improvement over the old. "I'm ashamed of you, Potiphar. Do you pretend to be an American, and not give way willingly to the march of improvement?" Potiphar knew that she wanted a palace but that it would turn out to be a "puny copy of a bad model." He called the new house "a cabinet maker's warehouse containing French wall paper, chairs like thrones from a Gothic cathedral," samplings of every classic pattern plus "new forms of furniture added to keep ahead of the neighbors," and "clustered without taste or feeling or reason."

Mrs. Potiphar had little regard for authenticity in her liberal borrowings from styles and artifacts of older civilizations. She took what interested her, or impressed her, or what she thought would impress others. She simply wanted her house to be the latest thing.

Once settled in their polyglot mansion, Mrs. Potiphar decided she wanted a coat of arms, uniforms for the servants, and a coachman who she believed must be fat and named James. When Mr. Potiphar objected to her plans on the grounds that such trappings were not consonant with the principles of democracy, she insisted that she was entitled to them because she was, after all, "lineally descended from one of those two brothers who came over in some of those old times, in some of those old ships, and settled in some of those old places somewhere." Mrs. Potiphar had a compulsion to imitate the styles and social practices of the past because she had no particular taste of her own and no other guide. She had only a vague idea of what a palace should look like. It did not matter that her imitation was a poor one. There were, after all, no palaces next door with which hers might be directly compared.

The various forms of cultural shorthand and mixing which Mrs. Potiphar enjoyed may be seen today in all parts of the country. Not only our homes but motels, restaurants, fast-food outlets, shopping centers, bars, furniture showrooms, and department stores display our penchant for imitation and cultural fakery. The most extravagant cluster of examples are the Las Vegas hotels, built, according to one architectural description, in Hollywood Orgasmic and Niemeyer Moorish.

In the more common cases, motels and restaurants have a fondness for harking back to older days of castles and coaching inns when life was supposedly an endless feast attended by comely wenches. The trappings for this illusion are supplied largely through the wonders of vacuum-formed polyurethane. Its ability to form moldings of simulated ornamental detail has unleashed upon the traveler Renaissance architecture, medieval furniture, beamed ceilings, and decorative wall plaques of knights, stagecoaches, and crossed lances. Even the new army barracks, designed by an Oklahoma City firm to attract men into the service, have plastic coats of arms on the doors.

Every hamburger stand, coffee shop, and motel lobby may now pretend to be a South Seas retreat, a Western ranch, a castle, or an olde tavern. A motel I stayed in recently was typical of those places that wish to appear "antique" but have little regard for logic or consistency. The motel's restaurant was called the Four Flames (I don't know why) and was divided into the Surf Room and the Turf Room. The two rooms were separated by imitation wooden beams, chipped here and there to appear hand-hewn and arranged vertically to form a modern room divider. Wall decorations included a colonial tavern sign from Vermont, a Germanic-looking coat of arms (the usual plastic imitation of a wood carving), some reproductions of old English sailing prints, and a plaque with fleurs-de-lis. I couldn't identify the furniture other than to say it bore some resemblance to that of the Middle Ages.

In the corridor leading to two cocktail lounges, called the El Cid and La Casa rooms, were two reproductions of paintings of dark-eyed señoritas with fans, sufficient to add a Spanish flavor. There was a night club in the basement made to resemble a cave. The walls were draped with gauze soaked in plaster of paris. The motel and restaurant were located in the mountains of North Carolina, but none of the decorative elements (except the cave) seemed to relate either to one another or to the locale or history of the state. In their odd juxtaposition of centuries and nationalities, the only unifying element was that the décor was more or less consistently "antique." It mattered not whether it was authentic in its imitation.

A New England motel I saw later had a general colonial theme in its décor, but with Florentine furniture in the bedrooms and a plastic palm tree next to the wooden Indian in the lobby. The swimming pool was done in a kind of Roman-bath style.

The same kind of casual mixtures are typical of Hollywood period movies. In the recent production of *The Great Gatsby*, Nick wears his hair long and thick instead of in the patent-leather plastered-down style of the twenties, and Daisy's hat is not the tight-fitting cloche but has a wide brim in keeping with current style. The distant past is represented by "any kind of vaguely old-time stuff whatsoever" and the "exact same costume trappings will do for all tavern brawls from the twelfth century through the eighteenth..."[1]

Cultural ransacking puts all design motifs into commercially eager and historically casual hands. One of my favorite examples of this is a "Morocco Nine Foot Fireplace, Bar, Stereo, Curio-Cabinet" that was offered by one furniture manufacturer. It was also available in other styles, referred to as "Mecca" and "Torrero."[2] The salesman told me that Torrero had "something to do with bullfights." When I assured him that it did not (and even if it did, bullfights have nothing to do with fireplace-bar-stereo-curio-cabinets), he said it was "probably—oh, a made-up word." Actually, a *torrero* is a lighthouse keeper or one who stands watch in a tower. The cabinet bore a faint resemblance to Spanish design, although it might have been called Mediterranean, North African, or Moroccan with as much validity.

Similar vague allusions are commonly the design antecedents of lamps, beds, cars, appliances, and fabrics. A furniture salesman told a Senate committee that the word "traditional," applied so frequently to furniture design, was meant to allow the customer to imagine whatever tradition he pleased.

Much of what we manufacture, build, or design imitates

or is named after something else. We see our automobiles as Thunderbirds, Falcons, Barracudas, Furies, Darts and Mustangs. Ford has a series of cars named after Spanish towns, although it is difficult to tell exactly what illusions they held in naming an automobile such as the Granada for a town whose only connection with automobiles is that it is overrun with Spanish Fiats. Chrysler, perhaps imitating Ford, recently marketed a "totally new car" called the Cordoba.

The U.S. Plywood Company makes wood-veneer and simulated-wood-grain wall paneling which is widely used in homes, particularly in basements, and in offices as a reasonably attractive and inexpensive way of finishing and decorating a room. In the naming of the simulated wood grains, imitation is laminated to imitation. The simulated teak grain was called Early Spring Cougar. The simulated pecan grain was called Early Spring Otter, and the darker simulated teak was called Early Spring Fox. To distinguish a color by naming it after an early spring animal, according to a curator I spoke to at the American Museum of Natural History, had no factual basis, but was "poetic." We can only guess at what dreams and images are tapped by these meaningless names.

Rugs, which come in a variety of colors and textures, are also given fanciful names that are nondescriptive but seek to draw on our Potiphar-like fund of vague memories and half-remembered impressions. What would a rug called Concert look like? The company that made this rug later changed the name, by some unknown logic, to Trumpet. Another of their carpets named McKinley was renamed Grand Teton and sold along with another carpet called Adirondack. What is the difference? And where are places like Saxon Park, Oxford Place, Cherry Hill, and Imperial House, and why do they have carpets named after them?

In Europe, athletic teams are generally named after the city, town, or athletic club they represent. In America, athletic teams are given additional designations. Some present a ferocious image (Lions, Tigers, Bears), while others seem either passive or neutral (Cardinals, Saints, Browns).

It is reasonable to ask why we regularly try to make a thing look or sound like something else. Why do we have Polynesian-colonial-Western restaurants, pirate-ship bars, Early American television sets, polyurethane Renaissance furniture, plastic palm trees, stained-glass linoleum patterns, and such juvenile and overwrought imagery for names of automobiles, room furnishings, and athletic teams?

The answer lies buried in our past and in the fact that we are a transplanted culture and that our image of the world and our selves is split between the two worlds that created us. The American settlers left behind all the material creations of their original culture and civilization—the villages and walled cities with their ancient cathedrals, houses, streets, bridges, and works of art. From this they came to a land of trees, sky, water, and open spaces, bearing no marks of civilization with which they could identify.

At first the settlers copied the old forms whether or not they applied to their new situation. Usually a mutation of form resulted, fitting the new and imitating the old. Many architectural features that had developed in England and the continent were followed in America in spite of the crucial differences in the climate and building materials of the colonies. In Virginia the first churches were "brick approximations of medieval prototypes from 'home'" complete with "vestigal buttresses."[3] The New Englanders imitated the thatched roofs of Old England until the practice had to be banned by law after it became evident that the drier and hotter American summers made grass, straw, and reed roofs extremely flammable.

In Britain, where timber was scarce, diagonal braces were used in house construction to give the soundest structure with the least wood. Amid an abundance of lumber, the diagonal brace was used in the colonies out of custom and taste. It is found as a decorative motif today, not as a reminder of the scarcity of wood but as a referent to some of those old times. A current example is to be found in a Lakewood, New Jersey, development house named the Shasta, apparently after Mount Shasta, a volcanic peak in California's Cascade Range. It features a "cathedral-ceiling living room, California patio kitchen with an L-shaped counter area," according to a press release, and was born out of European scarcity and Western extravagance. This kind of styling, loosely called "California" or "contemporary," is typified but a slope roof, high ceilings, and picture windows. When Larwin, a California developer, began what they called a "total community" on Staten Island with these airy dwellings, residents discovered they were almost impossible to heat in the cold New York winters.

The second-story overhang, a familiar sight in European houses, was copied in New World construction not out of reason, but purely as imitation. The overhang did not, as popular myth has it, provide a vantage point for fighting Indians, but was a custom in crowded medieval towns for the purpose of increasing floor space. The overhang is still popular with developers.

The rail fence was the first original thing built in the New World by Americans, borrowed from neither the Indians nor the Old Country. The most original buildings were the meeting houses of the Puritans and the Quakers, there being no European originals to imitate.[4] The log cabin, that humble dwelling thought to be the typical Early American house, was the work of Scandinavian immigrants and much too late for the Thanksgiving scene beloved of artists illustrating the November page on the calendar.

The separation of the colonists from their European origins and the challenges of the New World made it necessary to depend on new forms rather than on old ones, though the old was often discernible in the new. At first the memories of the Old World were fresh. Those builders of medieval churches in colonial Virginia had seen medieval churches in Europe. But as the generations descended, society's collective memory grew dim and was further confused by the constant influx of new arrivals with different origins and memories.

Culture, like language, has a spirit, a rhythm, and it is difficult to jump from one to another without a certain disharmony. Only rarely has a borrowed and imitated style blossomed in alien soil, but it does happen. Art Deco, to take a notable example, sprang to life at the Paris exposition of 1925, but it seemed destined for America because of its streamlined, rocketlike style, which suggested speed and "swift forward movement."[5] It left its impact on everything from toasters and lamps to the Chrysler Building.

Constant borrowings—"looting" might be a better word— from other cultures causes a deep alteration of the images the individual carries in his mind's eye and against which he compares everything from physical objects to personal behavior and value systems.

In America, the consistency and integrity of a style has ceased to exist. Heirs to the cultural traditions of all of Europe, we use a kind of ethnic shorthand in which it is possible for a fireplace with a built-in bar and hi-fi set to be named after what someone thought was a bullfighter. It is possible to mix German, English, Spanish, and colonial American decorations in one room without making visitors laugh, because they don't notice.

Like Mrs. Potiphar, who threw together disparate styles and periods, we find nothing strange about mixed motifs. One of the largest collections of Americana was assembled in Greenfield Village, Dearborn, Michigan, by Henry Ford and housed in a colonial Georgian building put together out of partial reproductions of Independence Hall, Congress Hall, and the old City Hall of Philadelphia. Ford had "something of everything" in his museum, from junk to rare antiques, displayed with no sense of order. A collection of dolls, for example, was put next to steam locomotives. One of his biographers called it a "hodge-podge . . . an Old Curiosity Shop, magnified 10,000 fold. It's disunity is as remarkable as the musical fare . . . [which] during [his] 46th Wed-

ding Anniversary called for Ave Maria, The Lord is My
Shepherd, The Last Round-Up, What a Friend We Have in
Jesus, and Happy Times Are Here Again."[6]

While Ford was an avid collector, he was not particular
about authenticity. He restored a schoolhouse in Sudsbury,
Massachusetts, which he insisted was the one memorialized
by the child's poem "Mary Had a Little Lamb." An old
woman who convinced him that she was the little girl in
the poem became known as Ford's Mary. He also found,
through his agents, a cottage in Pittsburgh he believed was
the birthplace of Stephen Foster. Thorough documentation
of his error by Foster's descendants was to no avail. Ford
carted the house off for his Greenfield Village restoration,
where it now stands, somewhat incongruously, next to a
reconstruction of Edison's Menlo Park laboratory.

Americans tend to enjoy a reproduction as well as an
original and have little sense of appropriateness in either.
When Glenn Turner, the man who made millions selling
musk oil, built a castle for himself in Florida (complete
with a bomb shelter), it could be dismissed as the excesses
of a poor boy who got rich. But when a resort developer in
Mecosta, Michigan, built a castle out of poured concrete,
you have to wonder what imitation castles mean to people.
A fake castle in Michigan has as little reason for existing as
does the London Bridge on an artificial lake in Arizona.

A Myrtle Beach real estate man offered to buy the Eiffel
Tower so he could move it to the coast of South Carolina,
and the French thought he was kidding. Perhaps they did
not realize than many Americans would have gone to see
it there as cheerfully as they would in Paris, counting themselves lucky to have saved the air fare.

While the landmarks of history are carted about or recon-
structed, the bulldozers and wrecking crews have been
busy. City after city, said Ada Louise Huxtable, "wrote
'good riddance' across the map to its past."[7] In city after
city, where centuries-old oak trees have been cut down to
widen streets, truck routes have destroyed quiet neighbor-

hoods. Hamburger restaurants rise on the site of historic houses, and brownstones become parking lots.

The physical remains to the past must be portable to survive, for only in this way can they remove themselves from the path of progress. The most trivial items can attract money. An Atlanta man claimed to have earned $10 million exhibiting the car in which Bonnie and Clyde were killed. And in January 1973, Hitler's 770-K Mercedes was auctioned off in Scottsdale, Arizona (not far from London Bridge), for $153,000; the new owner will display it in his Pennsylvania amusement park. When the Garden City Hotel on Long Island was demolished, a Levittown man bought the furniture from the room in which Charles Lindbergh had stayed on the night before he made his solo flight to Paris. And it was recently announced that the General Services Administration had acquired the instruments which the Parkland Hospital doctors in Dallas had used in the Trauma Room to try to save the life of President Kennedy.

Our indifference to the past, except in its more bizarre manifestations, is creating an American landscape that is as disordered as Mr. Ford's museum or Mrs. Potiphar's house. Our sense of the past is important, and its destruction will have serious consequences. Gore Vidal once made the remark that we become what we seem to be. In other words, appearances are the forerunners of what will be. If the appearance of the country makes it seem that we have no sense of authenticity, or history, and that we are driven by a compulsion to imitate, can we ever feel like "real" people in an atmosphere that is increasingly artificial?

"It seems just yesterday <u>we</u> were the Pepsi generation."

7

The
Generation
as Model Year

American nouveaumania, with its insatiable appetite for change and novelty, has in the last fifty years increasingly demanded that human beings, like manufactured goods, have model years. This urge has been largely satisfied by the excessive labeling of generations.

The *American College Dictionary* calls a generation a "whole body of individuals born at the same time," a definition that left journalists and social scientists at liberty to discover a new wave of Americans as often as they pleased. Sociologists agree, however, that the duration of a generation is about thirty years, or the time it takes for the father to be replaced by the son. The generation has been classically considered to be the embodiment of the spirit of the times or the age. This means that there should be about three "ages" in a century.

But in the first sixty years of this century, the United

States has, with characteristic excess, experienced four "ages": the age of the moguls and tycoons that preceeded World War I, the Roaring Twenties, the proletarian thirties, and following the transition of World War II, the age of suburbia and the organization man. In 1960, sociologist Bennett Berger, counting the beat generation and sensing the imminent birth of yet another generation, remarked that the United States seemed to have an age and a generation about every ten years, a rate which has since increased.[1]

David Riesman put the blame for the tendency to define generations on the mass media. The labeling of generations, he said, had been "speeded up in recent years by the enormous industry of the mass media which must constantly find new ideas to purvey, and which have short-circuited the traditional filtering down of ideas from academic and intellectual centers. We can now follow an interpretation of the suburbs from an article in the *American Journal of Sociology* to an article in *Harper's* to a best selling book to an article in *Life* or a TV drama—all in the matter of a couple of years—much in the way in which . . . a beat generation' [is] imitated almost before [it] exists."[2]

The attempt to name the generation and characterize the "age" became both a media and an intellectual pastime. Nor were young people entirely passive in the process. They rapidly acquired the notion that they ought to be completely different from the "older" generation, and willingly cooperated in the game of presenting a steady succession of new models for the satisfaction of their own egos and the entertainment of their elders.

In 1946 the youth of the day were called the New Lost Generation. Five years later they were called a generation of aesthetes. Also in 1951, a Texas college professor said they were without responses, and the term "silent generation" became widely used. The beat generation entered the national consciousness in 1952, and in 1953 William Styron, after rejecting "scared," called American youth a "waiting generation." A year later *Life* magazine said they were the luckiest generation. The 1961 generation was called both explosive and cool. In 1962 *Life* invented a generation out of young people who had already achieved unusual success in their fields. The "takeover generation," as they were called, consisted of only a hundred people, mostly white males. A swarm of photographers—I among them—was dispatched across the country to photograph this powerful mini-generation. They appeared in a foldout spread, each person about twice the size of a commemorative postage stamp.

The students and young people of the sixties really needed no such contrivance. They could boast of a record-breaking production of new generations (or life styles, as they were more popularly known), to fire the journalistic imagination. Reporters flocked to the campuses and discovered a new generation every year or two. Civil rights workers, hippies, yippies, show-biz radicals, revolutionaries, and various religious converts appeared and disappeared in rapid succession. The media called the 1963 generation tense, and in 1964 it was New Lost again. In 1965 it was the generation under the gun, and the youth of 1966 were open, restless, and rebellious. In 1969, they were called cheated and unsilent. In 1970 they were without fathers, and in 1971, romantic. For adults, the sixties were swinging.

In 1972, one young man told a newspaper reporter that he had been "into drugs," then had joined a commune, and ate only organic food. Now he was "into Zen" after going from other involvements ranging from stealing to loving Jesus. At nineteen he was, like many of his fellow students, a generation freak dabbling in life styles.

As the seventies wore on, the production of generations slowed down. The media grew impatient and one had the feeling that if the next generation did not appear, it would have to be invented. The press and television began to speak of the young with thinly veiled displeasure. The campuses were quiet, they reported. Apathy had returned. The September issue of *Esquire*, usually devoted to the college

scene, was given to other material in 1973. The editors felt
that the kids weren't doing anything and put them on notice
that they were expected to "go bite a dog or something."

James S. Kunen, author of *The Strawberry Statement*,
disparaged the inactivity of the students of 1973 and even
declared that his own class of '70 was copping out, not a
new generation after all.[3] A college freshman—class of
'78—acutely aware that his class was not leading a wave of
change, reflected that "It was a little disheartening to find
yourself drinking your parents' liquor and listening to
their old records."[4] Most students manage to see their own
activites as a progression. A New York University student
told a reporter, "More people have turned to drinking
because they have already experienced drugs and are ready
to move on to another level."[5] Presumably a higher one.

Many of the "old ways" are returning to campus. The
prom, dropped in many colleges in the late sixties because
it was old-fashioned, came back in 1974, complete with
waltzes and fox trots. Students also returned to the library
and grades were again important, giving a *New York Times*
reporter first sight of what may be called the new generation. He found that students were filled with such anxiety
over grades that they developed a neurosis and a tend--
ency not only to cheat but to sabotage the work of other
students seen as competition. There were no comparative
figures to substantiate the existence of a cheating or a
grade-grind generation, but it was interesting that the reporter found a new trend in an age-old college problem.[6]

While labels are only broad generalizations and are often
put forth frivolously, the label that sticks to a generation
is important to those to whom it applies, for, as Berger
noted, the successful designation determines who is behind
the times, who is "with it," and who is ahead of the times.
One's sense of esteem may be damaged by a derogatory
designation. The youth of the fifties, for example, were hurt
by the silent-generation label so firmly applied to them,
and they displayed a vague sense of apology and guilt.

When I was a freshman in 1953, a worldly senior told me of the glorious days in the late forties when the veterans had been on campus, of how great it was then, and how dead it was now. Veterans were supposed to have been a terrifically energetic generation of students, hellbent on success, a phenomenon attributed to the maturing effects of their wartime experiences. The football team even had a winning season. "There's no school spirit now," he said, "not like there used to be."

I recall quite vividly a political-science class in which the professor, with a trace of disdain, explained to us that in other countries, particularly in Latin America, students played an active role in influencing their governments, sometimes by setting things on fire and overturning cars. We're guilty, I thought, of not committing unlawful acts! I tried in vain to imagine the students I knew rioting in the streets, fitting us into grainy newsreels in my mind.

The only things the students of the Silent Generation burned were prep rally bonfires. And once a year fraternity pledges, bloated with beer and Dad's money, chopped a perfectly good piano to pieces and burned it in the middle of their lawn. It was in antics like this, along with snowball fights with the campus police, and panty raids in the spring, that the students of the fifties made their noise. We were trapped between the glory of the past and the hazy but persistent demands of the present, facing an unpalatable future outlined by William Whyte in *The Organization Man*, published the year before I graduated. We were expected to furnish the nation with a new generation more wonderful than the previous one—new and better, as it were. *Send me men to match my cars and appliances.*

Richard Lingeman, writing about his memories and experiences in the fifties, put his finger on the source of much of the prevalent guilt and regret. "We had," he observed, "no new life style to purvey . . ."[7]

The invention of new generations hardly contributes to the individual's sense of value. Berger pointed out that "a firm identity seems to manifest itself as pigheadedness and a stable one as stubborn rigidity."[8] Thus the American, already undermined by swiftly changing versions of humanity, is criticized if he possesses the qualities he needs most. To be excluded from the current age is to be obsolete, no matter how "into it" or "with it" an individual might feel. If he insists on trying to stay in style, he will look like that over-fifty executive in the *New Yorker* cartoon introducing himself to a young lady at a party as "part of an exciting new breed of old men."

Our emphasis on generations implies obsolescence of the older models. Unfortunately, Americans lack a sure sense of self-identity with which to counter the depreciating effects of this kind of rapid devaluation.

Our uncertainty is the creator of the stereotype of the American abroad, bellowing and ranting in hotels and restaurants over the illogic of the currency, the inadequacy of the bathrooms, the absence of fluent speakers of English. It is the picture of a man in utter confusion whose violent reactions arise from his having no sure sense of himself. He hides his internal incongruity in his blustering.

Humorists have always enjoyed the meeting of rustic Americans with the careful traditions of England and the Continent. The threat that the American feels among the people of older civilizations comes from the feeling of inauthenticity he experiences when confronted with a present linked coherently with the past. Henry James said that American society was "thinly composed," a condition he contrasted with the "thick, indubitable *thereness*" of English society.

The American lacks the sure grip that is the property of the past-connected man. Stephen Spender wrote that "Americans fear the European past" and added with equal insight that "Europeans fear the American future." History, even our own, is treated here as a pile of wornout events. (I encountered a Kansas couple in a French department store. They had just finished an extensive tour and I asked them how they had liked it. "Well," said he, "they just showed us a lot of history.")

Our separation from the past adds an American flavor to modern man's sense of alienation. We celebrate our freedom from the restraints of the Old World without connecting it with our malaise in the New World.

During the expatriot era of the twenties, when the rootlessness of our national life became unbearable to many, they sought to withdraw from their own society, back to Europe, seeking their culture's lost childhood. Though many artists and writers felt cut off, they remained abroad, seeking some living tissue to replace a vital part of their heritage misplaced in the transfer of European civilization to America.

Today we still feel the impulse to go look at Europe, to go "home" to places where we have never been, looking at familiar sights we have never seen. Our minds are filled with incomplete images of things past, half-formed visions we strain to see, hoping to find guidance in what we brusquely dismiss as "just a lot of history."

Our shaky identity structure is exploited by advertisers through the use of testimonials in which some well-known figure endorses a product or service, such as Joe DiMaggio's promotion of a savings bank, and the packaging of Mark Spitz. The association of a product with a clearly defined public image can have meaning only when our private image of ourselves is hazy.

In the absence of fully developed models of behavior or style, Americans are easily manipulated by those who claim to know the rules and can definitely state what is correct. The English are frequently held up as models, or as experts on taste. Americans are not quite sure of what "English" style really is, yet many carry the image of themselves as distantly associated with England and subject to that influence. There is, for example, an after-shave lotion called British John Bull, bottled by a New York firm. There is no relation between this product and any quality or suggestion of character remotely British, nor is there any way for such a relationship to exist. The lotion was assigned a nationality with the safe assurance that few would question its citizenship. To what people is such a product directed? Does John Bull need to be identified as British? And what exactly, in the manufacturer's opinion, does a Briton smell like?

The English obsession with what is "proper" is not easy for Americans to understand. When an Englishman says "That is a proper hat," he means that is exactly what a hat should be and no other will serve. Americans can't really think this way because we carry too many images of what a hat might be without being sure of which one, if any, is correct.

It is our confusion over identity, the absence of reliable models, that has given American character its malleability so plainly displayed in our readiness to be persuaded and manipulated by people who are only famous, to be overly respectful or hostile to other cultures that are only older, or to be repeatedly and variously marked by labels which are only rough generalizations.

Woman getting a "make-over"
at Amy Greene's cosmetic consultation center
at Henri Bendel in New York.

8

Fashion
as
New Self

"Apparel," said Shakespeare, "oft proclaims the man." In America, we have developed the notion that frequent changes in apparel and grooming proclaim as many new personalities, all released from the soul of a single individual. This idea became quite popular during the nineteen-sixties and early seventies when clothing styles underwent a series of radical changes that distorted the relationship between clothes and the man—or the woman—and created an illusion of eternal youth and constant rebirth through the packaging of the self. Fashion, in other words, got mixed up with inner being.

Prior to this time, clothing styles were largely indicators of status. Their dissemination followed a familiar and predictable course. Diana Vreeland, former editor of *Vogue* magazine, called fashion a social contract, "a group agreement as to what the new ideal should be." The agreement

*67

was made between the designers, who continually proposed something new, and the acceptance of those ideas by "the society that counts."[1]

Once this select group had adopted a new style, it filtered down to the masses, usually in a diluted form, until it became a cliché of dress, ready to be displaced by the next arrival. By this aristocratic but orderly system, the personal appearance of men and women all over America was constantly altered by styles adopted in the halls of the mighty and salons of the wealthy.

The fashions accepted by the trend setters were not arbitrary and random offerings of remote designers, but were, Ms. Vreeland tells us, reflections of the age, or in the words of designer Calvin Klein, "very today."

When the development of twentieth-century technology opened the vistas of mankind, the designers freed women from their corsets and bustles and gave them looser garments of brilliant color. And as the war approached in 1939, styles changed to "hard chic"—masculine suits with squared-off and padded shoulders.

In the last fifteen years, the increased tendency to discover new ages and generations brought confusion to fashion trends just as the events of those years brought confusion to our social and political life. The decade of the sixties was filled with tumultuous events and much shouting. Placards, chants, and slogans reduced discussion and debate to the size of a bumper sticker. People, as if attempting to achieve the same kind of instant communication, began to dress out their beliefs. Abbie Hoffman's American-flag shirt and his long, frizzy, uncombed hair were supposed to be an extension of his political views, a sign of his consciousness. He meant his attire to be a visible social comment, and as observers of the scene reported, "a costume is an idea, not a look—it's part of the person wearing it, not just a mere body wrapping."[2]

Young Americans came to equate fashion and personal style not only with their opinions but with their egos. Dress was an extension of self, but it did not simply proclaim *the* man, *the* self, in the Shakespearean sense, but *a* man, *a* self among many. Personal identity was not fixed, but changed with the times. A young writer about to turn thirty was able to reflect on four past selves in her precocious memoirs— headlined in a magazine article as "A last burp from the Pepsi Generation"—and view their obsolescence with the detachment of a historian reviewing ancient civilizations. First, she described her beatnik self of the fifties with "black turtleneck, beads, wraparound olive green skirt, sandals." She and her friends were constantly engaged in "deep talk," dividing the people of the world into two great categories, deep and shallow. Shallow included people who cared about clothes.

The second self came with Raga Rock, circa 1964. "The Beatles were in Kutra shirts and everybody knew how to say Maharishi Mahesh Yogi . . ." Third, there was the organic period. "I wore long skirts and cooked grains." The fourth self developed in 1969, when she "entered the women's Liberation Movement as a spy. I was going to write something snide. I came out a convert."[3]

This four-stage metamorphosis was roughly paralleled by the lives of many other young people, and in fact marks national epochs. Each phase, with its distinct costume, look, and concept of self, was as different from the preceding one as the butterfly from the caterpillar. American fashion trends were influenced by the individual's attempt to display an unlimited number of new selves; new editions of himself. Being in style is always being new, hence young, not just in appearance but—we are encouraged to believe—in essence.

As life style and clothes were seen as inextricably bound together, mass-produced garments were presented as expressions of the wearer's individuality. For example, *Mademoiselle* featured six young women who lived in different parts of New York City and who therefore each had a different life style. The article included photographs of each young lady in the clothes that matched her personality and her particular mode of living. The article made no attempt to explicate the relationship between the clothes, the life styles, and the young women, but the implication was that a relationship did exist, that it was not just a matter of taste that was reflected in the clothes, but some profound intuition.

The perception of new identities, personalities, and life styles expressed in the passing fashion carried the fable of the emperor's new clothes into another dimension. It was the emperor himself who was perceived to be new, and he acquired a new (imaginary) personality to go with his new (imaginary) clothes.

The style-as-self concept became a standard theme in the fashion magazines in the fifties. As a *Glamour* editor, my wife was an early writer on this theme. Later, as copy editor of *Simplicity* and *Modern Miss*, she set what may be a record in prescribing personal transformations, going from "Be a New You by Monday" (1960) to "A New You in Nine Hours" (1962).

Time has not dulled the editorial taste for transformation. The August 1975 issue of *Glamour* was headlined: "Make Me Over for My New Life."

No magazine has celebrated the link between fashion and self more lyrically than *Vogue*. It got particularly carried away in announcing the arrival of the "new prettiness," which it defined as a "whole new take on prettiness. Modern, upbeat, with a zing and style to it—the way we all want to feel in our clothes today, moving with ease through our lives, knowing we look well and not worrying about it—enjoying ourselves in fashion."[4] The *Vogue* editors never lingered over their definitions and seemed confident that their readers would be able to fill in the logical blanks, of which there were many.

Vogue is unrivaled in its ability to combine mental states, physical characteristics, and editorial omniscience. It saw clothes as being willed into existence by the woman wearing them. The clothes then transformed her as if they acted upon her physical and mental being in a manner perfectly divined, so to speak, by the editors.

Clothes, make-up, and fashion accessories were not promoted as simple objects but as humanized accomplishment of the self, ever new, ever up to date. This tendency to see people as coming in styles, with annual or seasonal model changes, is production-line American thinking.

The fashion magazines did not invent the renewable self, of course, but picked it up and amplified it. They played it back to their readers loud and clear, but with no hint of the problem of personal obsolescence which eventually plagues everyone in the land of the young.

The fashion system upon which the public depended for its supply of new "looks" has been under pressure in the last fifteen years. Prior to the early sixties, there were few extreme differences in the style of dress between age groups. Zoot suits and bobbysocks came and went, but in general, when young and old dressed up, they dressed alike. Even generational differences were minimized. In 1960 Bennett Berger said that the term "younger generation" was being stretched to include increasingly younger and older people. American children began dating, dancing, and drinking at a younger and younger age, while men in their late thirties still belonged to the "junior" chamber of commerce.[5]

When the generational war broke out in the mid-sixties, the stretching of the generation was abruptly halted and young people seized control of dress and style. A break occurred between age groups in fashion.

The admonition not to trust anyone over thirty also meant not to dress like anyone over thirty. With the advent of the hippies and yippies, campus styles changed radically. Young people let their hair grow and began to affect strange dress. The older end of the generational stretch found it difficult to dress, look, and act like the younger end. They were threatened with the social death of being outdated and of becoming one of life's discontinued models. Fashions ceased to filter down from the upper levels of society in quite so orderly a manner. They were now in the process of being forced up from below.

The English designers translated the generational split into mod fashions, and the American male, like Berenger in Ionesco's *Rhinoceros*, vacillated between standing his ground and joining the new herd, or as management consultant and business philosopher Peter Drucker called it, the Peacock Revolution. The blatant pretentiousness of some of the new fashions trapped many men between their desire to keep up with the times and a revulsion to dressing the part of a dandy. It was an interesting spectacle. Insurance salesmen, optometrists, agency account men, doctors, corporate executives, and accountants one had known for years as sober and conservative dressers suddenly appeared in flared pants, safari jackets, aviator glasses, square-toed shoes (later high heels), and wide ties that glowed in sunlight. Barbers reported business off 30 percent as men hot-combed their lengthening hair over their ears.

The disorder in the fashion system culminated in 1970 when the American woman rejected the midi skirt being touted by the establishment. The manufacturers, following the Paris showings, had invested heavily in the below-knee dress. The department stores in New York were loaded with the new midi look, but it didn't sell. The mass of women on whom manufacturers depended for sales and profits refused to ratify the social contract. Conditioned by the fashion anarchy of the previous decade, they refused to be pressured.

Having rejected the main line of the new fashion trend, they bought all the minor variations and novelties, causing Gloria Emerson of the *New York Times* to cry out in 1973 against the jumble of scarfs, pendants, bracelets, rings, oversized sunglasses, big-brimmed hats, platform shoes, wide pants, streaked hair, and shopping bags that made the New York woman a "mess." The "fashion industry cannot keep the trends coming fast enough. It is, after all, the very place

where the bullet belt and the fun-fun dogtag were launched
. . . If the idea is new, it must have merit. This is the reasoning and the results are often sad. . . . A wide range of synthetic materials makes us look wrinkled and limp . . .
Perhaps this is the natural punishment for caring so much
about keeping up with fashion. It is this eagerness, perhaps,
that makes so many women look so mindless and confused.

"Not trusting ourselves, we want to look alike. Someone
must tell us what to wear, what we should be pretending to
be."[6] Three years after the midi revolt, the field was in
disarray.

The impetus for these fashions did not originate with the
established designers, but came from the streets and campuses where the young congregated in protest and celebration. Their radical clothes were refined by the retail trade
with more conservative taste.

Clothes and hair styles were made to carry a heavy burden
of meaning. The older Americans who were somewhat reluc-
tantly influenced by the new styles found that they at least
had the virtue of being more changeable. If one's outward
style was accepted as a reflection of one's inner being, one
could display an unlimited number of new selves quite easily.

Since we believe that the transition from one style to
another is progress, the individual reviewing his past may
be chagrined to see his obsolete selves but is gratified that
his or her present self is the best, most improved self. The
new you is better than the old you. The "new prettiness" or
the "new beauty confidence" (another *Vogue* discovery) is
better than the old.

By way of contrast, the dilemma of the trapped housewife,
about which so much has been written, is that she has been
largely excluded from the game of changing self. The suburban housewife experienced no metamorphosis, but only
got a little older each year and more resentful of what she
was missing. There were no new life styles available to her.

One of the great traumas of American life is the sense of
being trapped in a style and shut off from the opportunity

to try on and exhibit a succession of selves. Maturity implies a settling of character, coming to terms with and accepting oneself. Nothing of the sort occurs in the evolution of American character, and the conformity demanded in the suburban-corporate world stands in conflict with our need for constant rebirth. The great feminist uprising generated its heat and power from the friction of this conflict and was ignited by stifled housewives trapped in their worn-out selves.

Widespread discontent in American society arises partly from the fact that growing old is equated with wearing out. It is difficult to see how any degree of well-being can be attained when people feel the constant need to feign eternal youth with a series of transparent selves which binds them to the destructive illusion that constant changes in style will buy them immortality on the installment plan.

"*I want you to know, gentlemen, that at this moment I feel
I have realized my full potential as a woman.*"

9

Womankind

A survey of high school girls in 1945 disclosed that most of the girls did not believe that the woman's place was in the home, as indeed it had ceased to be.[1] During the war, women had departed from their traditional domestic roles to become airplane mechanics, aviators, crane operators, bankers, bartenders, engineers, executives, diplomats, firemen (as they were called), financiers, forest rangers, garbage collectors, journalists, lawyers, lumberjacks, munition workers, physicians, radio operators, railway employees, riveters, scientists, sailors, soldiers, steelworkers, stewardesses, taxicab drivers, transportation workers, and welders. In July of 1944 there were about 18½ million women in the work force, or about 6½ million more than when the war began.

A wide variety of surveys, interviews, and editorial opinion suggested a deep split in public opinion about just what "place" women were going to have in the postwar world.

Men returning from the service were anxious to get women out of competitive jobs and many women were reluctant to leave. They had enjoyed being shipyard welders and liked the noise and excitement of running a business, chairing a committee, driving a cab, and the other jobs they had experienced. Many liked the economic independence and the wider world they had found through working outside the home. Rosie the Riveter and Wendy the Welder had been given much credit for helping to win the war. Afterward, however, they discovered they had gained a reputation for having become unfeminine. Robert Ruark, speaking to my news-writing class in college, said he built his career as a newspaper columnist by an attack on what American women had become. He compared them, unfavorably, with the European women, who were more graceful in the ways of pleasing a man.

In spite of some picketing and protesting by unemployed women demanding jobs, there was a general acceptance of the idea that in earning a living, men came first. After the war, when millions of women had given up their jobs, only about 500,000 were actively looking for employment. A *Fortune* survey made in 1945 revealed that 63 percent of the men, and even 57 percent of the women, interviewed believed that a woman should not be *allowed* to work if her husband earned enough to support the family.

Employers justified their preferences for hiring men by saying that women were susceptible to "fatigue and unhappiness," had a high rate of absenteeism, and quit work as soon as they got married. Margaret Barnard Pickel, Dean of Women at Columbia, denied none of these allegations in an article in the *New York Times*, but defended women by saying that they could not be expected to forgo marriage because of job considerations and that they would be "monsters of unnaturalness" if they did.[2]

Considering the many provocations, lack of understanding, and their massive expulsion from the work force, women had plenty of reason to revolt after the war. They did not.

To a large extent they left man's world and returned to woman's place.

A women's liberation movement could have started at almost any time since the war. But the discontent of women with wages, job discrimination, and male rule did not begin to be focused into a viable movement until 1961, when President Kennedy established the President's Commission on the Status of Women. In a paper on the genesis of the women's movement,[3] Jo Freeman said that is was the organizational structure and the communications network resulting from the creation of the commission that made the movement possible. The effect of the commission was to bring together politically active women in all fifty states, to uncover a wide range of evidence of the unequal treatment of women, and to generate an atmosphere of expectation that there would be something done about it.

While a movement needed a communications network, not just any network would do. "It must be a network that is cooptable to the new ideas of the incipient movement." And there had to be a crisis or a new idea to galvanize it.

In 1963 Betty Friedan's book *The Feminine Mystique* became a best seller and provided some of the galvanic force to touch off the movement. (Ms. Friedan had been one of those women who lost her job to the returning G.I.s.) In addition, the "sex" provision was added to the 1964 Civil Rights Act. When the Equal Employment Opportunity Commission failed to enforce the provision, it set in motion a chain of events that led to the formation of the National Organization for Women.

In 1967 a more militant group entered the movement. Many younger women, who had been evicted from the civil rights movement when the blacks decided to take over the major organizations, found themselves left as bystanders in the radical world. The draft-resistance movement was, actively speaking, for men only. As radicals without a cause, they were ready to discuss their own grievances against society, but no one was ready to listen. Freeman recounted

an incident in which the chairman of the National Conference for New Politics patted Shulamith Firestone on the
head and said, "Cool down, little girl. We have more important things to talk about than women's problems." After
similar experiences in other meetings, the new feminists
organized in a rage. They did not confine themselves to the
bread-and-butter issues of job discrimination and unequal
pay, but were soon involved in a rhetorical attempt to raze
the entire male-dominated social structure, proclaiming the
equality of clitoral masturbation, the myth of the vaginal
orgasm, and the superiority of the X chromosome.

The radical feminists carried the movement to the normal
American extremes. Everything would have to be changed.
Western literature (morally sexist), the language (structur-
ally sexist), the family (bastion of sexism), were in for
major overhauls and improved versions. Female self-suffi-
ciency would replace heterosexual relationships, and karate
would replace the last vestige of woman's need for man, as
well as her fear of him. Lesbianism became a political fad.
It was chic to be bisexual or at least to make a public
declaration of it as a new kind of revolutionary solidarity.

The plastic speculum was hailed as an instrument to free
women from their dependence on male chauvinist doctors
by allowing women to examine themselves. Marriage was
called a form of prostitution, and the care of small children
was considered an unreasonable oppression.

In 1969 and 1970 the faces of fuming women and the
sounds of their strident voices imprinted a new image of the
American woman on the public mind. Before the storm of
female anger, men ran for shelter, seeking refuge in the
natural order of things women accused them of inventing
for protection.

The women's movement made good copy, and the feminist's novel views on men, marriage, and society were
printed and broadcast as much for their entertainment value
as for their legitimate news value. The tremendous amount
of publicity given to the liberation movement hammered its

pronouncement into the consciousness of every American.

In the rush to freedom, a new conformity was imposed. One of the ideals of the movement was that women should have options, and should be free from enforced role playing. The movement developed with such speed that it soon turned on women themselves—those who had not changed as feminists thought they should, and who did not play the liberated role. The option to work, for example, became a requirement and compulsion. As Molly Haskell observed, "Such is the typically American lurch of the pendulum between extremes that the pursuit of a career is no longer just one of several options open to women, but mandatory."[4]

Diana Trilling, participating in a lib forum of the *American Scholar*, asked if the great majority of women were having their options narrowed by a privileged and determined few. "The right to be economically driven to the same degree that men are in our society may be our inalienable human right, but I do believe that women who are themselves professionally established and successful need to be cautious before demanding that right for the whole female population."[5]

Psychic stress is the price of change. Since the beginning of the liberation movement there has been a sharp rise in suicide among young women. In the under-twenty age group, suicide rates went from 0.4 to 8 per 100,000, and in the twenty-to-thirty group from 8 to 26. Sam Heilig, the director of the Los Angeles Suicide Prevention Center, said he had "never known a generation as interested in death as an experience, something you can pass through."[6]

Novelist Elizabeth Hardwick wrote: "If more young women are committing suicide it means they are cut off from the love of their fellow beings, or believe they are."[7] The separatism inherent in the movement urged the destruction of sustaining relationships that were supportive if not perfect. The movement produced adversary relationships in which females organized like trade unions against males.

Subsequently, older women formed a separate movement from the young and middle-aged, lesbians and bisexuals from the straights, blacks from whites.

Beneath the increasingly fragmented movement was buried the old liberal assumption of human perfectibility and American idealism, but missing in the perfection of the new sisterhood was the old concept of what used to be called brotherhood. As Elizabeth Janeway wrote, ". . . I think that the sooner women feel themselves as human being first, and female creatures afterward, the better off all of us (yes, humanity and yes, women) will be."[8]

Americans seem to be unable, as a mass society, to distinguish basic, fundamental change from superficial fads and fancies of the moment. The women's liberation movement offers an example of this. The public's attention in the last few years has been attracted by the more flamboyant issues of open marriage, the quality of orgasm, and the future of sex. In the meantime, a proposal of far-reaching social change was on its way to becoming a part of the Constitution. It has, at this writing, been passed in twenty-nine of the necessary thirty-eight states it needs to become the law of the land.

The Equal Rights Amendment reads like a platitudinous document containing nothing more than a democratic homily, but, in fact, it portends change in matters the public has neither heard nor debated nor fully understood. The heart of ERA is found in Section 1: "Equality of rights under the law shall not be denied or abridged by the United States or by any State on account of sex." That sounded fair to me, and when it appeared on the ballot in New York I voted for it. I had no idea, as I discovered later, what it meant.

Beneath that innocent statement of equality are legal implications for women and for society that the layman does not begin to suspect. An analysis in the *Yale Law Journal*[9] of the changes ERA would bring leaves little doubt that it would have a profound and surprising impact.

To begin with, jury law would be equalized. Women would no longer be excused from jury duty, as they are in New York State. Mothers of small children would not be excused unless the fathers were also exempted. Women would no longer be barred from jobs that require them to lift heavy weights, work long hours, or encounter hazards. Considerations of sex may be made only when it is a "bona fide occupational qualification." The details of "bfoq" have not been worked out, but in hiring practices, actors could be distinguished from actresses, for example.

Job security would have to be guaranteed to pregnant women, but mandatory leave during pregnancy would be abolished, unless it could be shown that the pregnancy caused a special job problem, which is considered unlikely.

A woman could not be legally required to change her name at marriage. The couple might, in some states, be required to have the same last name, but it might be his, hers, or a third name they both agree to. A man could not sue for divorce if his wife failed to follow him to a new job unless she could sue if her husband refused to follow her. The matter of who gets the children in a divorce would have to be determined in a "sex-neutral" way. Men could not be required to support their families. Alimony would be either done away with or made equally available to men.

If the military draft is resumed, women would have to be drafted along with men. One of the most controversial changes of ERA is that women could be required to fight in combat.

The Yale authors believed that separate schools and colleges for males and females could be allowed under ERA, as well as separate prison and restroom facilities, but Professor Paul A. Freund of Harvard is not so sure. If students or prisoners waive their rights of privacy, could separate institutions be maintained?[10]

Law professors may debate the ramifications of the amendment, but it is the general public that must decide whether or not it becomes a part of the Constitution. The decision should be based on an exact knowledge of its meaning. Our willingness to vote for change we do not, and cannot, predict is a remarkable fact.

Change often contains the illusion of progress, and equality the illusion of fairness. Unless we consider these prospects more carefully than is our usual custom, it is hard to tell whether the revolution they promise means advancement or turning in a circle.

As long as Americans are in motion they can be happy with the idea that they are getting somewhere. This showed up, for example, at the end of a panel discussion on women's liberation conducted by Michael Korda.

In summing it up, he asked, "You feel that we are moving and that perhaps we don't know where we are going?"

A voice: "But we're *moving* and that's what counts."[11]

The frenzied days of the movement are ending, and to use an old American expression, the novelty is wearing off. This was the mood found among a group of women interviewed at length by Catherin Breslin. All the women had been involved in the movement and were in various states of marriage, divorce, and affairs. "None of the women I spoke with," she concluded, "are saying we should go back to the old ways. But the new ways came along a few years ago with a heavy whiff of promise. A lot of women thought if they could find the heart to make some basic moves their lives would be fuller and better, not just different, and it turns out they are only different."[12]

Betty Friedan, herself wearied by the rapid changes, wrote recently of a need to pause: "After so much change in the sixties, we are breathless, tired after nearly ten years of naysaying."[13] It was becoming apparent, she said, to her and many other women, that they didn't want to change everything, after all.

*"To relate, respect, and yet be your own person,
until, perchance, affections erode?"*

10

Thoroughly Modern Marriage

When the GI's started coming home at the end of the War, the position of marriage and the family in American society never looked more secure. There were over 2¼ million marriages in 1946, a record that has yet to be equaled. The birth rate between 1945 and 1947 rose by nearly a third. The number of women in the work force dropped from the wartime high of 19½ million to 15½ million as Rosie the Riveter left the shipyard to raise a family.

Behind this wall of formidable statistics, however, the traditional family seemed to be coming unglued. In 1947 Dr. Carle Zimmerman, a Harvard sociologist, said that the "middle class family has reached its maximum demoralization (or will very soon) . . . If left alone, the family system will break up before the end of the century."[1] He noted, pointedly, that the Western family has collapsed twice before, about 300 B.C. in Greece and A.D. 300 in Rome.

The fact that the professor's dire prediction was publicized in the mass media (*Life* quoted him in 1947 and again in 1948) was a measure of the widespread concern for marriage in America. Many of the popular magazines were filled with thoughtful and anxious articles on the discussion of the family and suggestions for restoring it to health.

What caused the public's distress over the future of marriage and the family was the divorce rate. It had crept up during the war, and by 1946 it had doubled. Even though the divorce rate then declined for the next five years (and remained fairly constant until 1963, it continued to be cited as evidence that the family was in trouble and that the society was sick. The director of the American Institute of Family Relations, citing a study of 8,370 families, even declared that the divorced population was "biologically inferior."[2] However it was stated, divorce statistics meant that something had gone badly wrong.

As a result of the alarm over the jump in the divorce rate, most likely caused by a plethora of hasty wartime marriages, Americans began to think that the whole system was out of date and would have to be replaced. Della Cyrus, a social worker and wife of a Unitarian minister, wrote in the *Atlantic Monthly* that the family was better suited to the conditions of twelfth-century Europe than twentieth-century America. It had nothing to offer that could not be "obtained more cheaply and less painfully elsewhere." The sole explanation for the continued existence of the family was that it provided the only alternative to people who loved each other and wanted to have children. Since the "antiquated" family was no longer an exclusive source of food, shelter, security, entertainment, clothing, or sex, it put a heavy burden on love. The family was insecurely based on an emotional condition rather than a practical day-to-day necessity. Mrs. Cyrus felt that the "cooperative community" would be a viable substitute for the traditional family.[3] Alas, her new idea of a community never took hold.

Everyone had suggestions for altering the form of the traditional family, improving its behavior, and holding it together. In 1947 *Time* magazine carried Dean John Warren Day's eight-point program for saving the family. Dean Day, of Grace Cathedral, Topeka, Kansas, felt that there should be (1) a federal law to stop easy divorces, (2) large families, with less space between children, (3) better standards set by families depicted in movies, (4) no liquor ads, (5) sex education for high school students, (6) a conference with the county social hygiene clerk before filing for divorce, (7) a family pew so that the whole family could attend church together, and (8) daily Bible and prayer reading at home.[4] Dean Day's program was typical of much of the advice prevalent at the time and led the way toward the "togetherness" theme of the fifties that *McCall's* sniffed out of the still air and ran into the ground.

In 1948 a three-day conference in Washington, D.C., attended by nine hundred lay and professional people, focused on the family in a changing world. Many of the participants felt that the family simply wasn't changing fast enough to keep up with modern times. The director of a family institute in New York said "one of the major obstacles to better family living is this resistance to change." He felt Americans were too loyal to old ideas and living habits. Another participant urged that we "welcome ideas [about family life], welcome them in the spirit, we annually welcome the spring with its promise to renew life, new growth, and a world aroused from sleep."[5]

Changes in family life had indeed been stifled after the war. As one sociologist put it, "the pendulum swung back to neo-traditionalism" and family life was re-emphasized at the expense of educational and occupational achievements for women. A comparative study of values and attitudes of families in 1939 and 1952 found no evidence of change, with one prophetic exception: girls felt less of a "sense of obligation" to the family in 1952.[6]

David Riesman noted that in the late forties and early fifties an emphasis on the family began to appear in much of what he read. A man tended to refer to himself as having a wife and five children and a home in the suburbs, rather than stress his occupational achievements. The new shopping centers and supermarkets made grocery buying a family venture, and not infrequently an evening's entertainment.

The reassertion of the values of close family life in the fifties, the togetherness of it all, helped disguise the shortcomings of the blossoming suburban life, in particular. In 1960 Jhan and June Robbins wrote an article for *Redbook* that struck home in thousands of supposedly contented families across the land. "Why Young Mothers Feel Trapped" brought a flood of response, and for four years the magazine ran stories written on that subject by their women readers. Researchers at the Harvard School of Public Health found in their stories a "tone of loneliness and isolation."

The lid blew off in 1963 with the active beginnings of the women's movement, and the stresses of modern life that had wrenched at the family structure since the war could no longer be denied. The period of relative tranquillity was over.

The kind of change people had only talked about in the forties became a reality in the sixties. The cooperative community appeared in myriad forms of new domesticity: collective families, communes, marital partnerships, nonmarriage, contract marriage (complete with precise details and schedules for bedmaking, nights off, and alimony payments), single-parent families, and many other variations suggested a relief from the tiresome monotony of the nuclear family. Alvin Toffler, in 1970, tickled the nation's fancy in *Future Shock* with visionary predictions of a host of future unions such as the temporary marriage, professional parenthood, postretirement childbearing, corporate families, geriatric group marriages, homosexual family units, polygamy, and what will they think of next. *Life* magazine, having worried about the survival of the family for so long, gave way and ran a special issue on marriage experiments in 1972, describing the innovations in glowing terms and colorful photographs, a sure sign that such outrageous deviations had become respectable.

The death of the family became an established fact in America. The stresses on modern life had exerted a fatal pull on the familial alliance, bound as it was by tenuous passion and the accident of birth. As a provider of food, clothing, and shelter it had been supplanted by mass distribution and affluence. As a provider of sexual gratification, it had been supplanted by a new morality. "The family," wrote Robert Ardrey, "that final cooperative unit in the war against want, disintegrates for lack of function."[7] Once considered an essential human formation, man's natural social unit, the New Anthropology told us that the family was not a part of man's past until about ten thousand years ago and that it probably would not be a part of his future.

According to the senior demographer at the U.S. Census Bureau, the number of unmarried couples living together went up 820 percent in the last decade and in 1974 the marriage rate dropped for the first time since 1958. An article in a popular magazine offered practical suggestions for living together, such as the merits of separate medical insurance, décor, whether to live in her place or his, and ending with the firm advice to "stay away from married couples, especially those with children."[8] Nonmarriage, nonfamily living was seen as a newer and grander social experiment, an expression of modern independence. Detractors saw it as a lack of commitment, and marriage counselors said the unmarried had the same problems as the married: "They come to us fighting like man and wife."[9]

Nowhere was there an encouraging word for the survival of the family. Professor Ray Birdwhistle, of the University of Pennsylvania's Annenberg School, called it a closed dyad, "like a snake eating its own tail."[10] The president of the American Association of Marriage Counselors said, a little more succinctly, "Marriage stinks."[11] The author of a recent book on marriage traveled all over America looking for a happy family and was unable to find one.[12]

Even television, the great provider of home entertainment, began to ignore traditional family life in its programing. In the fall of 1974 the number of new shows in which the leading characters were not married increased sharply: *Get Christie Love!, Lucas Tanner, Manhunter, Chico and the Man, Paul Sand, Kodiak, Kolchak, Nakia, That's My Mama, Police Woman, Paper Moon, Rockford Files, Movin' On,* and *Planet of the Apes.* These new shows were added to those already established: *Mary Tyler Moore, Cannon, Ironside, Kojak, Gunsmoke, Six Million Dollar Man, Sanford and Son,* and *Kung Fu.* The married-life shows that remained were either sticky-sweet like the Waltons and the Apples or sour to bitter like the Bunkers and Walter and Maude.

On panel shows, in documentaries, and interviews the demise of the family was a favorite topic. In a single month (June 1973) WNBC-TV ran forty-nine different programs on the family and its crises, dilemmas, and failures. A full-page newspaper advertisement for the programs called family members an "endangered species." It became natural to think the worst of family life. When Bill and Pat Loud agreed to permit a camera crew to move in with them, I doubt that anyone expected the result would be a celebration of conjugal felicity, but not even the director of the film could have guessed what a disaster the marriage was. The Louds were not supposed to represent the "average" family, but as the serialization of their dismal union unfolded, critics and commentators nodded in agreement that the filmmaker had really captured the essence of life in the American home.

The heavy writers, James Thurber once said, had gotten sex down and were breaking its arm. The same might be said of what current writers are doing to the family. Professional journals and books, as well as publications intended for larger audiences, are replete with references to the death, obsolescence, or the need for radical restructuring of family life.

The bias Americans have in their general approach to change is to want to see it happen. It is difficult to escape the impression that current writers who predict the passing of the traditional family and the growth of alternate structures are expressing their own wishes.

Most people still marry, and they marry in the traditional way and live in families of conventional structure. The wild proliferation of communal life styles rashly predicted to be the wave of the future seem to be fading out. Divorce is again on the rise, but this can't be used as conclusive evidence of the death of the family. Of course, divorce is emotionally difficult, rarely genial, and expensive. In the words of the song, it's cheaper to keep her. That so many marriages end in this unhappy state may be seen as evidence of the importance of marriage, since remarriage is often the goal of one or both of the former partners. Even when legal remarriage does not occur, divorced men and women find new partners with whom to re-establish a family.

Supporting the view that divorce is not a sign of the decline of marriage, Berger and Kellner wrote that "individuals in our society do not divorce because marriage has become unimportant to them, but because it has become so important that they have no tolerance for less than a completely successful marital arrangement . . ."[13]

If divorce is caused by failure to create the private world, then one's home life is a crucial determinate of the sense of self in an overwhelming world. The family has indeed suffered a loss of function, but has also acquired a new one. The person headed for marriage may easily miss the distinction because, as Michael Korda observed, "our culture has trained us to look for the wrong things in marriage: passion and sentiment, instead of affection and care: excitement, instead of stability: individual gratification instead of the difficult task of building a life together and creating a world of one's own making, however small."[14]

Nouveaumania supports the "wrong" parts of these comparisons—passion and sentiment, excitement and individual gratification. These misplaced concerns have led to the consumerization of marriage.

The general rise in prosperity since World War II has made shopping a major activity. As a nation of consumers, we are urged to get more out of life—enjoy more, taste more, see more, do more—by buying new products which advertisers promise will make us secure, popular, envied, powerful, and sensuous. A hair spray, toothpaste, under-arm deodorant, or automobile is imbued with the power to make miraculous transformations in its user. Disappointment inevitably results and the consumer moves on to the next product.

This is, essentially, the ritual of the hysteric. Anthony Storr, the British psychologist, described the hysteric as someone who, in addition to other things, has a tendency to project "phantasy images upon real people. Ultimately, the actual person fails to come up to the phantasy, so that the hysteric is perpetually disappointed and looks else-where for (his or) her ideal."[15]

When we approach marriage as consumers, we demand that it live up to the advertisements we carry in our mind, which, as Korda noted, feature all the wrong things. If the marriage gives us trouble, we treat it as a product we have purchased. One does not owe allegiance to products, but demands performance, like the women who are causing male impotence by suddenly demanding that the earth move for them as it did for Maria in *For Whom the Bell Tolls*.

With our consumeristic attitude, if a marriage doesn't work we want to get rid of it right away and find one that works better. Paul Vahanian, a psychologist and marriage counselor, said of marriage, "This thing can't work unless you develop a frustration tolerance, and the younger people, especially, have never developed it. They feel so alone, they want to merge quickly, and when it's a drag, they split. It's obsolescence applied to human relationships; use it, throw it away."[16]

Lawrence Fuchs wrote that the family is the chief means by which we are connected with our past. But because we tend to think "change means progress," we do not revere the family.[17]

Dr. Kenn Stryker-Rodda, commenting on a recent increase

in interest among many Americans in genealogy, felt that it reflected a new awareness of "the sense of rootlessness in American life . . ." Without family ties among generations the "sense of human continuity is lost, sad to say, and there is a growing realization that it's something precious."[18]

The throwaway marriage fits the throwaway mentality Toffler found we had developed to match our throwaway products. He predicted that people in the future will marry many times as they grow and change. One reporter found that a "divorce mystique" has already developed in which divorce is seen as "an achievement in growth and self-analysis."[19]

The throwaway marriage is a form of serial monogamy which has never been a feature of Western nations for very long. Fuchs pointed out that in Samoa, serial monogamy is the established custom, but marriages are arranged for family prestige. The husband and wife never develop that passionate devotion to each other that Americans feel is the essence of a good marriage. The Soviet Union experimented with serial monogamy and sexual freedom in the twenties and thirties and was forced to denounce divorce and promiscuity in the face of the social chaos that resulted. "Many husbands left their families; rates of vagrancy, crime and delinquency among the young skyrocketed." The Russians are now quite prudish, by American standards. Fuchs also cited the study of one hundred societies by J. D. Unwin in *Sex and Culture*. This study concluded that no civilization which ever allowed sexual freedom for more than one generation continued to exhibit any great amount of "cultural and social energy."[20]

Philip Slater, in *The Pursuit of Loneliness*, saw sexual scarcity, appropriately combined with "economic and ecological" conditions, as the prime moving force in the society.[21] The mechanism of sexual scarcity has been so powerful that we have sexualized even the most routine products in order to sell them, a phenomenon familiar to anyone who has ever watched a television commercial.

We have become very conscious of ecology in the last few years, being told, to our amazement, of the possibility that spray cans are destroying the vital ozone layer in the earth's upper atmosphere and that the chlorine in the drinking water may combine with other chemicals to cause cancer. We have learned that there is a balance in nature that we must respect. What I am suggesting, and what a study of history also suggests, is that there is a balance in human affairs, too.

The great value we place on change and innovation leads us to call for a restructuring of basic institutions such as marriage and family without fully considering the consequences. The idea that something new will suddenly appear to solve all our problems and relieve us of the pain of reflections leaves us in our strained but characteristic posture, inclined toward tomorrow and its wonders.

"Just look at this report card! 'Accommodates enrichment data, inputs easily, and is generally self-programming.'"

11

Automated
Childhood

A housewife with children may utilize all the small miracles proffered by commerce and industry to make her job manageable, and still feel that her time is wasted and that she is trapped.

Technology and affluence promised her an easy life, as carefree and joyous as a Coca-Cola commercial. Pop-up, push-button, pull-out, instant, ready-to-use, disposable, carefree, electronic, solid-state, automatic—with these adjectives she is told that her problems have all been solved. And so she wonders why her child remains an inconvenience in a society that believes in the easy and the instant. When rising expectations call for a carefree, uncomplicated life style, children are stubbornly resistant to simplification. Child care is time-consuming and restricting in a society in which time-saving and personal freedom are among the highest values.

Children pin a woman down by what Gael Greene, speaking at a feminist conference, called their "tyranny," and in deciding not to have any herself, she chose "the freedom of being selfish."[1]

Many women, like Ms. Greene, have come to view children as despots who keep them out of a newer and more exciting world. Ellen Peck, author of *The Baby Trap*, told the same gathering that motherhood had been sold to women like Geritol. Childless women were happier, she said, citing a survey which showed that two out of five women who had children wished they didn't.

Traditional work at home, while being constantly "improved" by the products of various manufacturers, is held in low esteem. What traps and isolates women is the outmoded concept of motherhood that casts the mother in a servant's role, left with all the needless drudgery, forced to expend enormous amounts of time in service to the more pressing needs of the rest of the family who assume that she has nothing better to do.

In America the idea that a process cannot be made more efficient, that it cannot be made to run on schedule, is unacceptable. Lin Yutang called efficiency, punctuality, and the desire for success and achievement the three great American vices.[2] The mother of small children has always found the first two of these sins hard to commit, though she might have indulged in the third in her daydreams.

The scheduled and efficient version of childhood, which has evolved over the years as the status of motherhood declined, can be called the automated childhood, or motherhood by management. A pattern has developed which is dominated by the theme of separation—by the withdrawal of the mother from the child or the interposing of artificial barriers between herself and her child, even to the extent of dulling her consciousness of his birth when it is possible to be both conscious and free of pain, withdrawing her breast as a source of nutrition and security, and in general removing herself from his life on all but such ritual occasions as the reading of bedtime stories or visits to the zoo. In this new state, the child causes less work, takes up less of his parent's time, and gets less close attention from her.

The automated childhood is launched by the expectant mother who cancels her appointment for the afternoon, has a double martini, and grabs a cab for the hospital. The hour of birth is determined by her or the doctor's convenience. Barbara Seaman wrote that her doctor was planning to induce labor when she was pregnant with her third child until a public health pediatrician warned her that the procedure was dangerous and shouldn't be used unless there was a good reason. She asked her doctor what the reason was. He said he was booked on a cruise.[3]

The number of induced births for reasons of convenience is not known. The practice, in the words of a medical textbook, is "frowned upon, because in large series, about one fetal death in 200 inductions has been directly attributed to the procedure."[4]

The newborn is extracted from an unconscious mother to whom the anesthetized birth is indeed automatic, a shot of scopolamine having induced amnesia. The use of scopolamine has declined in the last twenty years in favor of drugs that allow the mother a degree of awareness and participation during the delivery, but some form of oblivion is the rule.[5] A study made in 1965 of almost 20,000 deliveries showed that only 14.6 percent of the mothers received no anesthetic.[6]

Natural childbirth in America is still regarded as a risky and exotic undertaking. Mental conditioning and exercise lacks that automatic quality that guarantees painlessness. As a friend of mine put it, "When I go to the hospital I want to be knocked out. When I wake up I want the baby there."

Interest in Read's "natural" and Lamaze's "psychoprophylactic" methods of childbirth has been world-wide, but it seems hard for Americans to accept drugless birth as an instance of progress. What it implies for most people is a return to a primitive, pre-chloroform state of raw pain. It is ironic that the Lamaze method, based on Pavlovian psychology, is by all humane considerations for mother and child the most "advanced" way of giving birth. It is used in almost all normal births in France and the Soviet Union, and is predominant in many other countries, though it was almost unknown here until very recently.

The desire for the unconscious, painless, automatic birth is a part of what doctors called our "medicated society" in which "we have eradicated some of the pain and anxiety, but . . . more of the excitement and joy. Pregnancy, labor, and delivery are thought of as essentially a disease in the United States."[7]

Dr. Waldo L. Fielding, in ridiculing natural childbirth, said the United States was already "utilizing the best tools that science has to offer" and claimed that "this country can now boast one of the lowest maternal and fetal death rates in the world."[8] Actually, the maternal mortality rate is about 30 deaths for every 100,000 deliveries, which places the United States seventh on the standard health index. We have an infant mortality rate of nearly 22 per 1,000 live births, a rate that places the United States thirteenth as compared to the other industrialized nations.[9]

We have always accepted as a matter of faith that new products of technology would make life better. Drugs are the major consumables of medical technology, and women have used them freely. A study made in California in 1963 found that the average woman used more than three different drugs of various kinds while they were pregnant, and 4 percent of the group studied took over ten different drugs.[10] After the thalidomide episode, it is now understood that drugs are potentially dangerous, yet an excessive amount of drugs are used routinely during pregnancy and at childbirth.

Almost all drugs lodge principally in the brain, the heart, and the liver. Drugs administered to pregnant women also pass quickly into the fetus. Since its liver has not yet developed the capacity for producing enzymes needed to metabolize the drugs, concentrations build up in the vital organs causing allergic reactions, sickness, and even death.[11] Ester Conway and Yvonne Brackbill, in a study made for the Society for Research in Child Development, pointed out that "Despite these ominous facts, the use of delivery medication in the United States is commonplace, its absence a rarity."[12]

Researchers are now working on the problem of getting the embryonic liver to produce the enzymes needed to metabolize drugs, leaving for future study the problem of what happens to the liver when it is forced into early production and what happens to the system when enzymes not normally present begin to show up.

Dr. Watson A. Bowes, in reviewing the history of drugs administered to pregnant women, wrote that "the fetus is potentially at greater risk from well-intentioned medicaments than from the vicissitudes of pregnancy and delivery...

"It is surprising that the fetus has fared so well, subjected as it is to so many agents. But as physicians and the pharmaceutical industry develop ever more potent and complex ways of altering physiological and pathological states, the fetus is bound to be the inadvertant target of maternal therapy."[13]

In *Future Shock*, Alvin Toffler describes the "new birth technology" of tomorrow that would include "embryo implants, babies grown *in vitro*," and "the ability to walk into a 'babytorium' and actually purchase embryos..."[14] In spite of the brave new world overtone, this is the kind of medical virtuosity that appeals to our nouveaumania and could complete the development of motherhood by management, in which *all* direct ties between mother and child are severed in the name of progress.

Nothing better demonstrates the withdrawal of the mother than the decline of breast feeding. With the successful formulae for breast-milk substitutes, the physical needs of the newborn baby can be tended to by anyone. The baby can be fed a manufactured formula, now available in six-packs, with disposable nipple and bottle.

"Technical developments," Margaret Mead wrote, "have made it less necessary for the child to be continually in the care of some nurturing person.

"We have the baby carriage, the playpen, and the feeding bottle, and now the plastic bottle, which means that the child can be very early left alone with a thing, rather than a watchful person, without danger of broken glass and when the baby is separated from the mother because she no longer breast-feeds, it also can be separated in all these other ways. Paradoxically, this makes it easier to leave the child with another person, but also easier for that person to neglect it, leave it lying for hours in a crib, or listlessly waving a piece of old rag back and forth, or rocking or banging its head."[15]

Not all mothers who give up breast feeding do so out of choice. It is suspected that the use of anesthetics during birth is a part of the cause of a decline in breast feeding. The drugged, unresponsive baby who does not begin nursing soon after birth often has difficulties beginning later on.

Although the debate over the relative advantages of the bottle and the breast remains a topic for dispute, there is little in the way of conclusive clinical evidence that breast-fed babies are healthier or more secure emotionally than bottle-fed babies. One recent study concluded that there was "no significant differences which favored breast feeding rather than formula feeding for boys and girls." However, the study concluded that "boys who were nursed [i.e., by breast or bottle] for a long period ... by a warm mother tended to be relatively free from difficulties," while those who were nursed for a long period by a cold mother had the most pronounced maladjustment.[16] What seems to be the important element in the infant's life is not really the kind of milk it gets—although there are arguments for the nutritional superiority of breast milk as well as for its immunizational properties—but the presence of a warm and loving mother. Since there are no serious arguments for bottle milk but plenty of arguments for the baby's need of a regular and continuous relationship with its mother, it is pointless to pursue the debate over breast milk except to suggest, as Margaret Mead did, that the return to breast feeding might be a means of legitimating the working mother's right and need to be with her child, perhaps making it necessary for employers to give the mother time off during the day.

In the automated childhood, the mother-manager does not consider it necessary to spend a great deal of time with her child, but adopts the role of overseer and coordinator negotiating with others who provide the day-to-day, or minute-to-minute, care. She may find these arrangements complicated and time-consuming as the number of available nannies continues to decline and the number of working mothers entering the job market continues to increase. A recent *Times* article recounted the experiences of women who imported girls from the South, the Middle West, or from abroad. One woman, unable to find a governess, hired a "governor"—a young Frenchman who runs the house and the children. Another, unable to find a day-care center for her nine-month-old, quit her job to organize one of her own.[17]

Whether by choice or necessity, 42 percent of American mothers now work. As the absent mother takes on an executive position in the family and leaves the line functions to what is usually a succession of hired sitters, the relationship between mother and child becomes less intimate and maternal deprivation becomes an issue.

Some researchers have challenged the concept of maternal deprivation because of the vagueness of the term and the data supporting it. The Israeli exeprience with the kibbutz has been cited as evidence that the mother need not be with her children all the time. The kibbutz, however, is not analagous to the situation in most American homes. The kibbutz was formed as a colonization device to promote nationalism. And further study shows that all the problems, conflicts, and failures found in the United States are present in the kibbutz and that kibbutniks generally do not choose to bring up their own children in that style.[18]

The most authoritative studies of maternal deprivation have been done by John Bowlby. He has graphically described the typical symptoms:

"The initial phase, that of protest, may begin immediately or may be delayed; it lasts for a few hours to a week or more. During it the young child appears acutely distressed at having lost his mother and seeks to recapture her by the full exercise of his limited resources. He will often cry loudly, shake his cot, throw himself about, and look eagerly toward any sight or sound which might prove to be his missing mother. All his behaviour suggests strong expectation that she will return.

"During the phase of despair, which succeeds protest, the child's preoccupation with his missing mother is still evident, though his behaviour suggests increasing hopelessness ... He is withdrawn and inactive, makes no demands on people in the environment, and appears to be in a state of deep mourning."

The mother, on subsequent visits, is largely ignored by the baby. "Should he, as is usual, have the experience of becoming transiently attached to a series of nurses each of whom leaves and so repeats for him the experience of the original loss of his mother, he will in time act as if neither mothering nor contact with humans has much significance for him."[19]

The person who is "mother" to the baby is someone who, in Margaret Mead's words, is "intensely interested in him or her, who will spend endless hours responding and imitating, repeating sounds, noting nuances of expression, reinforcing new skills, bolstering self-confidence, and a sense of self."[20]

Nothing about contemporary American life encourages the

mother to provide this kind of sustained attention. Bringing up children has become a cottage industry in an era of automation, personally demanding at a time when the mother's attention seems designed to wander.

The issue of maternal deprivation is of major importance as the family is being divided between the home and the day-care center, the use of which is a manifestation of the decreased personal involvement of parents, mothers particularly, in the lives of their children.

Without adequately considering the consequences, we are pushing toward a society in which children will be institutionalized in the name of freedom and convenience. If this seems to be an overstatement, it should be realized that the Child Development program passed by the Congress in 1972, and vetoed by the President, would have established a network of day-care centers across the country, operating on a twenty-four-hour, seven-days-a-week basis. Although the funding of the initial program was $2 billion, the full estimated cost was $30 billion.

Institutional care would radically alter the relationship between the child and his parents, and consequently between the child and society. Visions of sparkling, well-staffed Child Development centers may have the aura of efficiency and convenience that Americans find so appealing, but results from other countries should be studied more carefully. In reviewing a government study of day-care centers in the Soviet Union, Hungary, East Germany, Czechoslovakia, Greece, Israel, and France, a *New York Times* writer reported that "The risks for many, though not all, children range from mild neuroses and developmental lags to serious maladjustments."[21]

The small child who spends most of his waking hours in a day-care center will undoubtedly be more the product of that center than of his own home. As a government manual on day-care operation put it: "The child care center operator is responsible for children . . . perhaps most important [for] their education and character development.

"When children spend six to ten hours a day in a child
care center, education and character development cannot
be left to themselves; whether they wish to or not, the
adults in the center affect the children and the influence of
the parents is diminished."[22]

Qualified day-care-center personnel is in short supply.
Margaret Mead called it almost nonexistent. One can imagine a vast network of government day-care centers run
with all the care and imagination of the post office.

The United States has been called child-centered, but the
gradual estrangement between parent and child suggests
otherwise. We like the idea of youth, of being young, and
keeping up with what young people are doing, but nobody
wants to be stuck with the kids any more.

The automated childhood carries the inherent assumption
that people can have their children and not get involved.
As we grow away from intense involvement toward a more
fractured and independent future, we take less personal
responsibility for the kinds of people our children become,
but ironically, we are very much influenced by what children do *en masse*. It is the detached and kaleidoscopic
world of the young that closely resembles the life adults
are reaching for, a world imagined to be new and free.

Children are given very little guidance by parents who
envy and imitate youth, and who are confused and unsure
of their own values and only too glad to relegate the task to
the teacher, the sitter, the minister (if there is one), Captain
Kangaroo, or in the guise of open-mindedness, let the child
find his own way, which he will do after a fashion.

The myth that Americans are child worshipers is a mis-
interpretation of the fact that we cling to our own childhood, mistaking our obsession with youth for a concern for
children. Nouveaumania is not only the love of new things
and new ways, but it is also the love of being, feeling, or
appearing ever young. It is difficult to accept the aging
process in a society that worships youth, that offers no

guides to growing old, no secure place for the aged. "Children" or "young people" are the only age categories that enjoy a high status in American society, but little firm direction. A study of Finnish and American children found that American children were more frequently influenced by their peers than by their parents. Even when parents have strong ideas about the behavior of their children, they find it difficult to express them. Dr. Haim Ginott, the child psychologist, remarked that "everything we do is done with hesitation. Even when in error, grandfather acted with certainty. Even when right, we act with doubt."[23]

A few years ago it would have been unthinkable for a daughter to bring home her boyfriend from college and expect to share her bed with him. When a group of parents were questioned about their attitudes on contemporary bundling, even those who were most scandalized found it hard to object without being defensive. None of the parents who opposed the sleep-in date were entirely secure in opposing it and many of them allowed it.

Lawrence Fuchs wrote that since the nineteenth century, Americans have viewed the young as "more important and virtuous than the old," whereas in every other major culture children have been regarded as "untrustworthy" until they were properly civilized by their elders. Following their children, Americans have created for themselves a permanent artificial childhood.[24] The *New York Times* commented in an editorial on the great extent to which adult fashions copied children's clothes. "The store windows this season are filled with pinafores and puffed sleeves, smocks and knee socks and Mary Janes and ponytails, tied with ribbons or held in place with a plastic rabbit. Colors are pale pink and sky blue and Easter-chick-yellow; prints show Mickey Mouse and lollipops. Kids are in."[25]

In the universities, dissertations were written on comic books, students watched kiddy television—*Howdy Doody*, *Buffalo Bob*, and *Sesame Street*—and at Yale, the most pop-

ular seminar was on children's literature. These were not
mere excursions into nostalgia and camp but an attempt at
"artificially reconstructing childhood."[26]

Catalogues filled with adult toys arrive in my mail with
unencouraged regularity. Stores like Hammacher Schlem-
mer in New York carry a complete line of gadgets to appeal
to the child within.

The return-to-childhood theme in advertising is not difficult to discern, particularly in airline travel and automobile
promotions. The American car is not a sex object. It's a big
toy, even if there were breasts on the Buick and a vagina
on the Edsel. The tail fin was designed by every boy in my
third-grade class, years ahead of Detroit, and tail-finned cars
were marketed about the time we came of age.

The most persistent symptom of our longing for lost childhood is the aversion to responsibility, noted in the frequent
use of phrases in which responsibility is something one is
"saddled with" or finds "heavy," implying that it is a joy-
less, unrewarding burden. It would be interesting to explore
the theme of running away in fiction, dominant in such
divergent works as *Huckleberry Finn*, *Rabbit Run*, and *The
Glass Menagerie.*

The aversion to the sticky business of bringing up children
is legitimized by the threat of overpopulation which has
substantially altered our feelings about what used to be
called blessed events. The sight of a crowded nursery evokes
the same environmental anxieties as a traffic jam on the
expressway.

Motherhood has become the occupation of the unskilled,
the trapped, or the deluded who "wallowed in the aesthetics
of it all—natural childbirth and nursing became a maternal
must. Like heavy-bellied ostriches, they grounded their heads
in the sands of motherhood, coming up for air only to say
how utterly happy and fulfilled they were."[27]

With a swiftness surprising even for America, motherhood
fell from its place as worshiped myth to a misdemeanor, for-

givable in the ignorant but punished by ridicule and pity among the educated.

For the sake of their children, American parents will spend anything but time, pay anything but attention, and give anything but guidance. Thus the child bears the brunt of our nouveaumania. At every stage of his life he is subjected to the fruits of our search for the better way, for convenience, for time-saving shortcuts.

There is a limit to the degree to which we can substitute efficiency for personal involvement. We become accustomed to reality once removed, as a processed, frozen dinner is an abstraction of food, and a credit card an abstraction of money, which is itself an abstraction of labor. Our reality is vicariously perceived. In this artificial style, we lose tolerance for things that require too much of us, that demand our whole being, and that accept no abstractions, no substitutes for being there and paying that kind of intense and careful attention which child care demands.

12

The French-Fried Connection

In 1945, as the end of the war drew near, newspapers and magazines began to feature articles on the imminent changes in what and how Americans in the future would eat. After the war, there would be a "dazzling assortment of new [food] products" and all would be "labor saving."[1] There would be greater variety in food than ever, and it would be easier to shop for, to store, to prepare, and to clean up after.

Novelty, variety, and convenience were clearly going to be more important than quality. The foods might, as advertised, taste "just as good" or "almost as good" as the traditional homemade products, but no one promised they would be better. And in fact, postwar foods featured improvements in everything but taste.

Everyone was in a hurry. For the last sixteen years there had been depression and war. Now suddenly peace and relative prosperity gave Americans renewed confidence in the future and a sense of urgency to get moving, to make up for lost time. The prospect of spending less time in the kitchen and the store had great appeal to consumers. A new food that was easier to prepare could be forgiven much in the way of price and flavor. Most of the new foods that were put on the market had the common characteristic of being ready to eat or nearly so.

Wartime food research had dealt with preservation, bulk reduction, and new food sources. As a result, in 1946, proposed or new consumer products were an assortment of dehydrated and anhydrous foods, canned meals, canned meats, (more than forty kinds had been developed for the Army), canned bread, irradiated meats and vegetables, electronically baked, nonmolding bread, treated and hybrid eggs that would resist spoilage, nonmelting butter, ground fish, dried eggs with pulverized shells, and dehydrated banana, skin and all. Menhaden, a bony fish normally used for making oil and fertilizer, was canned and renamed silver herring after a method was found for cooking the fish until the bones dissolved.

Instant potatoes, developed too late for Army use, went on the grocer's shelf along with instant cocoa, dehydrated soup and fruit, dry skim and whole milk. Food researchers succeeded in extracting the juice from vegetables and making a powder of the extract, thus reducing a bushel of spinach to a small capsule. Following this triumph, they experimented with rice husks, coffee grounds, and chicken bones as sources of new food extracts.

Freed from the demands of wartime, food producers unleashed on the public a continuous buffet of novelties: barbequed salmon, turkey sandwich spread, and pâté of smoked rainbow trout. The greatest impact on America's eating and shopping habits, however, came not from wartime food research, but from the popularization of frozen food. The quick-freeze process was developed by Clarence Birdseye in 1924. He found that when big portions of food were frozen, or when any amount of food was frozen slowly, large ice crystals formed in the tissues and destroyed the cellular structure. The food, when thawed, had a mushy texture. Small portions, Birdseye discovered, could be frozen rapidly without this unpleasant effect.

The marketing of frozen food, which began in 1929, was hampered by the lack of food freezers in the stores, but in 1938 the first all-frozen-food store opened in White Plains, New York. Others followed. Up until the outbreak of the war, consumption of frozen food grew at the rate of about 25 percent a year, although it remained a small part of total food consumption. During the war the processing of frozen food leveled off, but many people with victory gardens began to freeze the surplus of their harvest, either in their own home freezer, if they were among the few to own one, or in freezer plants where individual lockers could be rented.

After the war the food industry had high hopes for frozen food. While many of the postwar miracles turned out to be duds, the frosty cornucopia lived up to expectations, at least in terms of growth. By the middle of 1946 there were more than five hundred brands of frozen food on the market, and three hundred and fifty different foods had been frozen either commercially or experimentally.[2] In New York City there were more than forty retail stores devoted exclusively to dispensing frozen fare. Sales, as they say, boomed.

Article writers, imagining the future, wrote that frozen food was going to make the shopping trip a monthly or bimonthly excursion and the necessity to cook a thing of the past, leaving the housewife free to enjoy an endless holiday. Enthusiasts claimed that the finest cuisine of the world's best restaurants would be instantly available in the family freezer. And buying frozen food was going to be *fun*. It was going to be sold in "self-service dispensing units, just like playing the juke box."[3]

Frozen food came close to doing some of the things that were expected of it. The shopping trip became a weekly

ritual, and the housewife, if she never got that endless holiday, at least had less cooking to do.

Some pretty fair restaurant food did make it to the supermarket; but the most enduring change was that restaurants
themselves have become dispensers of frozen food and the
chef has become a "thawer-outer." None of the postwar
miracle foods had a greater leveling effect on the nation's
diet. The frozen dinner arrived as the embodiment of the
future and has remained as a threat to the survival of cooking as an art.

In our postwar eagerness we rushed to the future, em-
bracing everything we thought would speed us to the promised land of imagined golden tomorrows. We did not dream
of the pleasures of *haute cuisine*, but rather looked for
wonders in the miracle foods, the instantized, nonperisha-
able, overprocessed creations of the mass marketers. It was
our hunger to literally taste the future that has led to the
plastic quality of American food today.

To continue to produce new foods, an increasing amount
of what we eat must be heavily processed, doctored with
additives, or entirely manufactured. D-Zerta, a General Foods
whipped dessert mix, is made from hydrogenated vegetable
oils, whey solids, sodium caseinate, propylene glycol mono-
stearate, acetylated monoglycerides, hydrooxylated lecithin,
sodium carboxymethylcellulose, sodium saccharin, artificial
flavor, BHA, citric acid, and artificial color. While D-Zerta
may be an extreme case, and not exactly a basic food, the
use of a wide assortment of chemicals in food has become
common, and their use has created a $500 million industry.
There are now over 2,500 additives in commercial use as
anticaking agents, emulsifiers, preservatives, antioxidants,
humectants, glazing, firming, crisping, foaming, and release
agents, and solvents which are used for flavoring and color-
ing.[4] There is even a chemical made to improve the machin-
ability of bread.

Food additives create many of the characteristics thought
to be natural properties, such as the red color of hot dogs

which comes from various combinations of sodium erythor-bate, sodium nitrate, citric acid, ascorbic acid, and glucono delta lactone.

The most frequently used food additive is flavoring, neces-sary not only to inject taste into highly processed food but to give vent to the nouveaumania of the food industry, allowing such products as cheese twists that are hot-dog-flavored and potato chips that taste like pizza.

Shape and taste together can be manipulated to provide a wide assortment of food gimmicks. One company recently marketed a line of vegetables for children who "hate vege-tables." Carrots, green beans, corn, beets, and spinach were given the shape, flavor, and texture of French fries. And if you have wondered why onion rings are now all the same size and French fries the same length, it is because they begin as a "matrix mix," a slurry made of onion or potato. The mixture can be extruded into crescents, rings, shrimp shapes, nuggets, sticks, and logs. DCA Food Industries, which developed the process, was given the top honor in the 1971 Putnam Food Awards Program, sponsored by *Food Processing* magazine, for the fabricated onion ring. The company has also used the process to make apple-and-cinnamon rings; shrimp, ham-and-cheese, and clam rings; potato au gratin; and mushrooms. More new products are on the way.

Almost everything in the American diet is now partly arti-ficial in that it has been altered from its natural state by packaging and processing which to some degree, changes the shape or flavor or nutritional content of the food. The American taste for the artificial may be considered in its more literal sense in the food we eat.

Oranges appear as juice, -ade, or practically in name only as orange "drink" made of sugar, water, orange flavoring and little or no orange juice. One food critic suggested it should be called orange water. Such fruit drinks frequently contain a simulated citrus pulp to create what processors call "mouth feel."

Corn, on the rare occasions when it is found on the cob, is usually sold (often frozen) without the husk and packaged in Saran Wrap, its flavor lost. For many city people there may be no choice, but even when stores and roadside stands are laden with freshly picked corn, people still buy it packaged.

Tomatoes are likewise sealed in plastic, usually when they are still green, and taste like wood by the time they reach the table. The flavor and juiciness of a vine-ripened tomato is becoming a lost memory. In the American style of seeing all change as progress, a columnist in 1946 wrote approvingly that in the future, "exposed" produce "would become as old-fashioned as the cracker mill."[5] He also predicted that hamburger patties would be packaged in strips of cellophane and sold by the yard, an idea which seemed to delight him. It is perhaps symbolic of a growing disenchantment with "progress" that only dog food is sold that way today.

The roast beef, chicken, turkey, pork, and ham sold in most delicatessens and restaurants is sliced off from processed, reconstituted loaves, all of which taste the same. Fish, as sold in fast-food chains like McDonald's, is processed to be odorless and nearly tasteless. A processing-plant sales manager told reporter Mimi Sheraton, "We sell more fish now that it doesn't taste like fish."[6]

The traditional Thanksgiving turkey found in the frozen-food department has not escaped the forces of change. The new turkey has been "prodded to grow with feed additives, infused with artificial color to mask an unnatural paleness of flesh, injected with artificial flavor to make up for the natural flavor lost during forced growth, treated with phosphates to bind in water absorbed in a chilling tank, and in some cases, injected with a basting fluid."[7] One of the most widely advertised of the self-basting turkeys is Swift's Butterball Turkey, which contains no butter, but an artificial butter flavor.

The convenience and variety of processed food are frequently exaggerated. Frozen spinach, which shoppers will

buy even when fresh spinach is available, saves little time
or effort. The same is true with peas, corn, and other vegetables that have short cooking times. And no form of instant
potato can compete with the simplicity of a baked potato.
I have recently seen frozen chopped onions and chopped
peppers, and it is hard to imagine why anyone would buy
them. What is it like in a home where no one can chop up
an onion?

According to a cookbook I took down at random from the
shelf, there are seventy-two ways to prepare a potato besides
baking it. Processed potatoes are sold mainly as potato
chips, French fries, or in dry form for making mashed potatoes. A plethora of brands and minor innovations conceals
this poverty of selection. On a recent visit to a large supermarket, I paced off over 240 feet of shelves and display
racks on which there appeared to be an endless assortment
of snack foods, but setting aside minor differences, there
were only six basic products. Thus the consumer's demand
for novelty and variety is met with magnified differences in
a single product. A minor variation becomes a major theme,
keeping the itch for the new and different with limits that
can be easily scratched. It need hardly be pointed out that
the national taste for such "foods" is expensive, wasteful,
and of questionable benefit to our health.

I counted over twenty six different brands, sizes, flavors,
shapes, and thicknesses of potato chips, each proclaiming
some special feature. They came in little packages and in
large drums; in thick, thin, and wavy slices; in flavors of
barbeque, onion, cheese, garlic, and believe it or not, sour
cream. The Pennsylvania Dutch Style chips I bought owed
their claim to uniqueness to having been cooked in shorten-
ing instead of vegetable oil, although there was no difference
in taste. The newest chips were castings of perfectly uni-
form shape.

Sometime in the not too distant past, a person or persons
unknown put dried onion soup into sour cream and the
cocktail dip was born. Brittle potato chips often broke off

when plowed through the viscous mixture, and a thicker "dip chip" was soon added to the snack counter. After its initial success the dip languished in the monotony of its original form. It has now been revived, and I counted twelve different kinds of flavoring for the sour-cream dip.

Like the potato chip, the ancient pretzel, once bound to its unique contorted shape, has found new life on the supermarket shelf. By a rather loose count, I found twenty-four different kinds of pretzels.

The remaining space on the snack-food shelves was taken up by croutons (twelve kinds in the standard supermarket flavors of barbeque, onion, cheese, bacon and combinations thereof), cheese puffs, corn chips, and popcorn.

The same overproduction of new products that typifies the snack foods prevails in other parts of the store.

The creation of new food products has been greatly enhanced by the development of artificial foods. It has recently been announced that in the future all food can be entirely manufactured and that "texture, mouth feel, flavor and color can be *essentially* duplicated."[8] (Italics mine.) Texturized vegetable protein (TVP) will take over at least half of the processed-meat sales by 1980, according to some research estimates, and by 1985 Americans will spend $3.5 billion on TVP products.[9] Such "engineered foods," as they are also called in the trade, have been developed by a number of major food producers and are based on a formula of spun soy beans chemically manipulated to resemble a known food. Worthington Foods, for example, is now marketing synthetic breakfast sausages, patties, and slices of imitation ham. To overcome the unpleasant connotations of synthetic food, the new products are being sold under the "warm, homey sounding" name of Morningstar Farms. General Mills and Swift are expected to enter the market also.

Another group of imitation foods is being made from calcium algenate: imitation spaghetti and sauce, fabricated meatballs, artificial caviar, shrimp, soft-drink gels, fabricated cherries (without a stone, the promoters say), blueberries, onion rings, potato chips.

The labeling of artificial foods is sure to be a source of future controversy. In 1973 the Food and Drug Administration proposed that the word "imitation" appear on food labels only if "those foods are nutritionally inferior to their imitations. If a fabricated product were nutritionally equal to a natural food, the imitation could be marketed without specifically using the word."[10] As the number of TVP and other food products are added to the market, consumers will not likely want to lose the distinction between the artificial and the real.

Besides the engineered foods, the other great source of new food products will be the so-called convenience foods. Convenience and novelty, while not without benefit, are proven enemies of quality and nutrition. Every time food is "improved," it gives up flavor, freshness, and nutritional content and acquires another chemical. Improved food might be easier to prepare, but it more commonly means that it is easier to produce and market. Longer shelf life, less spoilage, ease of handling, shipping, and packaging mean more to the producer than to the consumer, who must eat the preservatives, pay for the packaging (which sometimes costs more than the contents), and accept a decline in quality.

One of the newest convenience foods to be introduced was Hamburger Helper, a mixture added to hamburger meat to make a dinner in one pot. The Kraft Foods version was a set of six dinner mixes packaged to look like books. Any one of these products will make a meal in twenty-five minutes that has a chemical odor and a "flat, synthetic and often objectionable" taste.[11]

Probably the ultimate in convenience is the food that comes in aerosol cans. Cheese spread in a pressurized can sells for 79 cents for 4-⅝ ounces, or $2.73 a pound. It is more expensive than most gourmet cheeses, and it tastes like glue. Aerosol dispensers are being used for catsup, syrups, salad dressings, and dessert toppings. A mixture of peanut butter and jelly has even been tried.

The road to instant coffee was paved with improvements for the sake of convenience. People used to make coffee by simply adding the freshly ground coffee to hot water and then stirring in a little cold water to settle the grounds. Coffee beans were roasted daily in neighborhood shops and ground at home just before use. In San Francisco around 1900, ground coffee was first sold in vacuum-packed cans. The new convenient canned coffee put the roasting shop out of business. Shortly afterward came the percolator, which somehow struck people as an easier way of making coffee. It wasn't, nor was the coffee as good.

As the quality of a cup of coffee fell, importers began to buy cheaper, inferior beans because no one could taste the difference. We were thus prepared for instant coffee.[12]

As novelty and convenience food dominated the family table, home cooking came to mean food that was unwrapped, mixed, or thawed at home. Food that was "almost as good" as the original was followed by food that was almost as good as the imitation. Standards were lost and as one woman sadly noted, "I never know what they mean when they say 'season to taste.' "[13]

It is predicted that in the future the American diet will consist entirely of convenience foods and that friends will drop in for dinner unannounced "because it will be understood that food won't require hours of preparation."[14]

The decline of taste in American food has occurred in a carnival atmosphere. The frivolous tone set by the emphasis on novelty is plainly visible in the supermarket setting. The bright flourescent lights, the Muzak, the banners and crepe paper that typically festoon the ceilings and counters suggest a continuous party as the shoppers browse through the rows of Giggle-Noodle soup, Krazy-Glazy Toaster pastries, and cans of Beef-O-Getti. The purpose of the party is to keep alive the spirit of adventure and fun conducive to the sale of such items as Green Giant frozen peanut-butter sandwiches, Boo Berry cereal (also Count Chocula and Franken Berry), Max-Pax coffee, and spray-can food.

Since the end of World War II we have eaten an increasing amount of processed food. In 1952 Americans had a per capita consumption of 102 pounds of potatoes a year, 8 pounds of which were processed, mostly as potato chips and French fries. In 1962 Americans ate 80 pounds per capita of fresh potatoes and 27 pounds of processed potatoes. This was the year the fast-food business began to grow. In 1970 we ate 119 pounds of potatoes, about 50-50 between fresh and processed. It is estimated that by 1980 we will be eating twice as many processed potatoes as fresh.[15]

Dr. Alexander M. Schmidt, head of the Food and Drug Administration, told *U.S. News & World Report* that about 49 percent of the food served at home is prepared elsewhere and that by 1980 about two thirds of it will be.[16]

Occasionally there are signs of incipient revolt against the increased consumption of processed food. According to sociologist David Berger, "There's a real middle-class movement away from frozen food to real cooking. Men are cooking elaborate things. Perhaps it's coming out of the barbeque phenomenon."[17] One frequently encounters references to the connection between the artificiality of modern life and the food we eat. For example, when the Parsons School of Design decided to make a bread cookbook instead of an ordinary yearbook, one of the students said, "A lot of my friends are baking. It's a change of attitude to doing things that are real."[18]

As long as our diet is influenced by nouveaumania, we can expect more and more plastic food. In most countries in Western and Eastern Europe with which I am at all familiar, there is a stability in the national diets. Cuisines have changed some with the passage of time, but they remain remarkably stable, and by our standards, boring in

their lack of variety. In America we still like to think the
food we eat today will be obsolete tomorrow, that a change
in diet is progress.

"Who can predict the excitement, the intrigue of food of
the future?" asks a Durkee ad in *Food Processing* magazine.
"Most of them haven't even been dreamed of yet. But already we're figuring out how they'll be made . . . Before
tomorrow's foods can be made and tasted, they must be
dreamed. You do the dreaming. We'll do the rest."

Or is it a nightmare?

DIRECTIONS FOR USE

If COFFEE RICH is frozen when received, defrost in your refrigerator until completely fluid. When defrosted, keep under refrigeration. Shake gently before using.

INGREDIENTS

Water, Corn Syrup Solids, Vegetable Fat, Vegetable Protein, Polyglycerol Esters of Fatty Acids, Polysorbate 60, Dipotassium Phosphate, Disodium Phosphate, Carrageenan, Artificial Color (Beta Carotene).

13

Terminal
Nouveaumania

Since the end of World War II, then, the American diet has changed from the traditional three squares to a smörgåsbord of convenience foods. According to a recent study, the average person makes about twenty "food contacts" a day as he consumes "oreos, peanut butter, Crisco, TV dinners, cake mix, macaroni and cheese, Pepsi and Coke, pizzas, Jell-O, hamburgers, Rice-a-Roni, Spaghetti-O's, pork and beans, Heinz catchup and instant coffee."[1]

Such a diet consists largely of "empty calories," fats, and sugar. Middle-class Americans can no longer speak of the undernourished poor, because studies show that the majority of all Americans are nutritionally deprived.

Continuous snacking on junk and novelty foods is responsible for the poor nutritional health of the nation's teen-agers. Mary Goodwin, a nutritionist, told the Washington, D.C., Heart Association that the American life style

made a balanced diet almost impossible. Teen-agers in particular eat "from vending machines, at ballparks and movies
and are offered nothing but junk food with no nutrition . . .
snacking has become a way of life."[2] It is perhaps not surprising that a Cleveland couple held their wedding reception
at a drive-in hamburger stand and drank toasts with strawberry milkshakes.

The hamburger, because of its dominance of the fast-food
menu, has become the symbol of American cuisine. Rubbery
though the extruded mass-produced patty may be, high in
fat, carbohydrates, calories, and sodium, and a poor taste
for children to acquire, it is probably our most frequently
consumed food.

The success of the hamburger is attributable in large
measure to its presentation as a new food under the guise
of Big Boy, Big Mac, the Whopper, the Quarter Pounder, the
Steerburger, and Big Whitey (whose name had to be
changed after it acquired racial overtones). Pizzas, fish and
chips, and fried chicken have enjoyed the same kind of ongoing rejuvenation. Americans have eaten fast food in such
enormous quantities that in Japan and Europe young people
are eager to try it, "ready to be convinced it's good."[3] (A
friend of mine told me that when a McDonald's opened on
the Champs Elysées, posters had been prepared in which
the Big Mac was translated as Le Grand Mac. At the last
minute the French printer had to tell the company that Le
Grand Mac meant chief pimp in a whorehouse. How well the
hamburger itself will translate into other cultures remains
to be seen.)

The urges of nouveaumania that have made a success of
mass-produced food have maintained a constant demand
for new food and there has been no lack of enterprise among
manufacturers to provide it. To do so, they have resorted to
a variety of food fads, gimmicks, and artificial manipula-
tions, though gourmets and scientists may object on the
grounds of taste and safety. Julia Child and James Beard
have firmly denounced the soyburger, and a group of inter-

national scientists called for a "working party" to plan ways to combat the long-range effects of "internal pollution" caused by ingestion of synthetic substances in food.

The most controversial of the substances are the chemical additives, particularly the nitrates. These chemicals are used as preservatives in all cured meat—ham, bacon, pastrami, corned beef, and most sausages—and also to control color, flavor, and texture. Nitrates in high doses can be converted by the stomach into nitrosamines, which, according to Dr. William Lijinsky, a cancer researcher at the Oak Ridge National Laboratory in Tennessee, can cause cancer and severe anemia.[4]

Artificial colors and flavors widely used in such "kid foods" as rainbow-colored cereals, flavored potato chips, soft drinks, and drink mixes may cause hyperactivity in children. This charge was made by Dr. Ben F. Feingold of the Kaiser-Permanente Medical Center in San Francisco in his recent book *Why Your Child Is Hyperactive*.[5] A preliminary study by the National Institute of Education has supported Dr. Feingold's charges. The Food and Drug Administration, finally responding, persuaded the Food Research Institute of the University of Wisconsin to make a controlled study. When the results are reported, the FDA plans to study which additives are harmful and if "different children are affected by different additives," a process that could take years. In the meantime the agency has no plans to warn parents about the possible dangers.[6]

While the specific effects of a single food additive on the human system may be tested in the laboratory, the long-range effects of the many different additives, in combination with one another and with diet pills, tranquilizers, and antibiotics, is more difficult to assay. Food additives which are harmless by themselves may form toxic compounds once ingested. For example, studies indicate that the sugar substitute Aspartame could cause brain damage in children when it was combined with monosodium glutamate, a common "flavor enhancer" widely used in processed foods.[7]

Engineered foods have become of special concern because of their heavy use of chemical additives. At a Washington hearing on Department of Agriculture regulations on "alternate foods," Joan Gussow, a nutrition instructor at Columbia University, testified along with thirty other nutritionists and ten citizen groups against the increased use of "mock foods" in the school-lunch program. "This is part of the general trend toward more synthetic food in the American diet, but we have no evidence on the long range implications of a synthetic diet." Susanne Vaupel of Food Research and Action Center in New York City called the 25 million children in America's schools a captive market for junk food sold under the guise of good nutrition. She noted that Secretary of Agriculture Earl L. Butz had been on the board of directors of the Ralston Purina Company, a manufacturer of textured vegetable products, and that the previous Secretary of Agriculture is now on Purina's board. E. J. Hekman, administrator of the Food and Nutrition Service of the Department of Agriculture, was formerly president of Keebler & Co., maker of fortified cakes.[8]

Such cross-pollenization of government and industry is not unusual, nor does it prove any kind of collusion, but it does raise questions about the objectivity of government decisions to use engineered foods of dubious value for school-lunch programs. Furthermore, in 1974 a study found that government agencies were "poorly organized to pursue a national food and nutrition policy" and that they lacked a national goal. No one was in charge.[9]

As American consumption of processed foods—canned, frozen, dried, and engineered—increases each year, the question of their nutritive value becomes more important. As consumers we do not see what goes into their preparation and must often guess at their content. Only recently has the FDA issued any rules about labeling for nutritional content —calories, protein, fats, carbohydrates, vitamins—but many companies are balking and compliance is still voluntary.

One of the difficulties that processors face in nutritional

labeling is that vitamin content declines as the food ages. A statement of nutritional value at the time the food is processed would be off by 10 to 20 percent in a few months. The inevitable deterioration of preserved food is not a question they wish to draw public attention to.

Another problem for processors in complete labeling is that the figures can be embarrassing. Fruit juices such as Hi-C and Hawaiian Punch turn out to contain only about 10 percent fruit juice and very little vitamin C. Canned chili con carne, according to *Consumer Reports*, contained as much as 17 percent fat, with the meat portion consisting mostly of cartilage and head meats, (cartilage protein is worthless and the protein in head meat is incomplete). Any complete labeling of contents would have to include the Defect Action Level, which specifies, among other things, how many insect fragments and rodent hairs are allowable. In the chili study, samples from all 49 products contained some filth. "We found insect parts and short rodent hairs which are all too characteristic of ground spices. We found whole insects, which suggests poor storage conditions for ingredients. And we found long rodent hairs, which usually indicates contamination at the processing plant."[10] Since the high processing temperatures sterilize the contents, the infestation is only an "esthetic" problem.

One of the great nutritional hoaxes of all time is the familiar white "enriched" bread, heavily promoted as essential in the diet of growing children. Dr. Roger J. Williams, a biochemist and nutritionalist, found the bread so void of nutrition that it can't keep mice alive. In Dr. Williams' experiment, mice were fed nothing but white bread; they all died of malnutrition. He wished to call attention to the larger issue of the "apathy," if not the antagonism, exhibited by classical medical education toward nutrition, which he feels is the basic cause of the backwardness of the baking industry.[11]

Many nutritionists suspect that there is a relationship between the decline in quality of our national diet and the

decline in our level of health. Death from malignant neo-
plasms rose from 64 per 100,000 in 1900 to 150 per 100,000
in 1960. Heart disease, directly related to diet, has risen
from the ranks to become the leading cause of death. Adelle
Davis, noting the increase in a wide range of diseases, wrote
that "the tremendous increase in ill-health has paralleled
the ever-mounting consumption of sweets, refined foods, and
soft drinks, and the corresponding decreased use of fresh
vegetables, whole grain breads and cereals, legumes, and
potatoes."[12]

The relationship between the decline in health and the
rise of artificial, overrefined processed foods is established
through the testimony of vivisected malnourished mice and
guinea pigs, autopsies of young soldiers that reveal signs of
arteriosclerosis as a general condition, the increasing inci-
dence of cancer among the young, and the lack of stamina
in schoolchildren.

Exercise also plays its part in good health. The elevator,
escalator, moving sidewalk, and the automobile have not
made it impossible to walk, but to borrow from an old joke,
thank God, we don't have to. Our lack of exercise and our
taste for the "new and improved" foods have impaired our
health. To remedy this, we look to new and improved medi-
cines, medical inventions, and surgical techniques to repair
the damage. Our handling of sugar is illustrative of this
process.

Six generations ago, when the per capita consumption of
sugar was fifteen pounds a year, coronary disease was rare.
This fact was cited by Dr. Robert C. Atkins in defense of
cyclamate. Dr. Atkins, in a newspaper advertisement, urged
the government to reconsider its ban on the artificial sweet-
ener on the grounds that it had never been definitely proven
harmful to humans, while sugar was implicated in our most
prevalent diseases: heart disease, diabetes, obesity, hypo-
glycemia, peptic ulcers, and others.[13] The Sugar Institute
frequently runs advertisements praising sugar as an energy

food, although our present levels of consumption are known to be harmful.

Sugar is the favorite ingredient of food processors. John Hess, former food editor of the *New York Times*, observed that they put it into bread, catsup, cereal, baby food and almost everything, and then take it out of desserts.

Cereal manufacturers, while denying any connection between sugar-coated breakfast cereal and poor nutrition or tooth decay, have refused to disclose the amount of sugar in their products. According to the Center for Science in the Public Interest, six of the twenty-eight brands tested were 40 percent sugar, and Quaker Oats King Vitamin was half sugar. The Center, along with the American Public Health Association and the American Society for Preventive Dentistry has asked for a 10 percent limit (by weight) on the sugar content of presweetened cereals.[14]

American children are addicted to sugar not only because their breakfast comes coated with it, but because they are encouraged to eat enormous amounts of candy and sweets of all kinds. Hundreds of millions of dollars are spent each year on advertising ($81 million on breakfast food alone) aimed directly at children, sometimes with a little help from the government.

When the Continental Baking Co. was asked by the Department of Agriculture to produce a high-protein food to be given out at school to poor children who got no breakfast at home, the company made a small cake called Twinkie which "tasted like a devil's food cupcake with a sweet, creamy filling."[15] Twinkie is currently being fed to half a million children in St. Louis, Memphis, Little Rock, and Atlanta. Nutritionists denounced Twinkie as junk food, and Dr. Louis Ebersold of the American Dental Association pointed out that it was laden with plaque- and decay-producing material and should not be served to children who had no chance to brush their teeth after eating it. (A Hoffman-LaRoche ad in *Food Processing* magazine suggested that the "breakfast of tomorrow" might be a fortified straw-berry-ice-cream cone.)

In the light of what is known about the harmful effects of high sugar consumption, it seems strange that we continue to feed sugar to our children and to ourselves. The search continues in the laboratories for a safe substitute for sugar, but the simplest solution is to cut down on the amount of sugar we consume each day. We should stop giving sweets to children as a reward for good behavior, get the lollipops out of the pediatricians office, the sugar out of baby food, Twinkie out of the schools, and either get the sweets commercials off television or the television away from the children. (It is ironic that the Children's Medical Center of New York Fund is currently soliciting funds with a Lollipop Campaign in which money is collected by "Lollipop people" who give out the candy with "Thanks for giving" written on it.)

The Senate Select Committee on Nutrition in April of 1973 made public a 1967 report that showed a "very high" rate of cavities among children who ate presweetened cereal. And another study showed that rats fed Tang, the orange-flavored drink, had significantly more cavities than those given distilled water.[16] Fluoridated water, tooth coatings, mechanical brushes, and irrigating systems are not the answers to the dental problems of people who live on sweets, and yet, typically, we expect mechanical devices and chemicals to solve our problems.

The excess consumption of sugar has not been linked to heart disease with the cold certainty demanded by the clinician, but the evidence is compelling.

Although heart attacks account for over half of all deaths in the United States, doctors find that the preventive measures of diet and exercise are not well received by most patients. Dr. William B. Kannel, a leading researcher in heart disease, says that the patient feels he can get "that kind of advice from his mother-in-law. What he wants is a pill to counteract the effects of his bad habits."[17] The heart patient is more comforted by medical technology than sensible advice. He is impressed by the number of intensive-care units for coronary patients. These new units grew from none in 1961 to over 2,700 in 1971. Over half the people who have a heart attack each year die within the hour. Of the 650,000 Americans who die every year from heart attacks, about one third die within a month. An ounce of prevention is worth a lot of intensive-care units.

After the excitement of the first heart transplants came the realization that life spans were lengthened only by a matter of months, an achievement which could have been accomplished with less pain and suffering by beginning earlier in life to follow a moderate program of exercise, to quit smoking, and to reduce intake of sugar and high-cholesterol foods.

It is typical of the nouveaumania in our culture that routine preventive medicine is less interesting than medical technique. In contrast to the great public interest in "miracle" drugs, exotic surgery, new hospital construction, and medical computers, preventive medicine is almost completely ignored.

We live on a junk-food diet, nutritionally deficient and cholesterol-high, and feel fortunate that we live in time of medical progress. Yet the life expectancy of adults has improved very little since the nineteen-twenties.[18] The one radical difference in the mortality rate of 1900 and 1974 is the decline in the number of deaths in the first year of life.[19] The only medical advances that have significantly lengthened the life span are the vaccines used against childhood diseases. None of the surgical showboating or miracle drugs has contributed to average longevity. (According to John Knowles, president of the Rockefeller Foundation and former director of Massachusetts General Hospital, there is an excess of general surgeons and neurosurgeons in the United States.)

A long and healthy life for the greatest number of people should be the first in any list of medical priorities, but only

a few researchers have studied the healthy. Dr. Alexander
Leaf, a Harvard Medical School professor and chief of medical services at Massachusetts General Hospital, therefore,
made an unusual trip when he visited three areas in the
world where men live the longest: Vilcabamba, Equador;
the principality of Hunza in Kashmir; and the highlands of
Soviet Georgia. While his survey was by no means exhaus-
tive, Professor Leaf found that diet, exercise, and social use-
fulness were the common characteristics of fit and active
men whose ages ranged from about eighty to a hundred and
twenty-one. Their intake of fat was from 40 to 60 grams a
day in Georgia, and as low as 20 grams in Vilcabamba.
Americans consume an average of 157 grams of fat a day.
Daily caloric intake in Equador and Soviet Georgia was
1,700–1,900 and 1,200, respectively. The American's "average
daily intake of 3,300 calories, including substantial quantities of fat, is excessive and conducive neither to optimal
health nor to longevity."[20]

In October 1972 the Carnegie Commission issued a report
calling for more doctors, more nurses, more hospitals, more
medical and dental schools, although the United States now
spends more on medical care than any other nation. The
cost of medical care is fast reaching the breaking point, but
the general health levels in the United States are inferior to
most other industrialized nations and are getting worse.
The president of the National Health Council called the
health system in the United States so ineffective that another $10 billion "would not substantially improve it."[21]

Prevention is not our style, but it may be forced on us
because we cannot afford the cost of the technology that
has already been developed to treat the most prevalent
diseases. According to Dr. Knowles, all attempts to hold
down medical costs in the future will fail. "Kidney dialysis,
transplantation surgery, artificial organs, heart-lung pumps
—these and more to come would inevitably consume an
ever increasing share of the national resources, and all
attempts to reduce or hold down cost of medical care would

be doomed to failure. The heart transplant is the medical equivalent of landing men on the moon."[22]

The morality of how medical funds are allocated will be the subject of many future debates. It will have to be determined whether we should continue to build a complex medical technology or broaden our concerns to major health problems viewed as a part of how we live. A poor diet and a lack of exercise are almost socially enforced. Unless we are able to practice preventive medicine, which begins with a proper diet and a modest amount of exercise, medical services in the future will have to be rationed to an even greater extent that they are now. Disease is now recognized when the patient sees the doctor, but its origins must be located in our daily lives. Medical "progress," as well as the food we eat, should be what keeps us healthy, rather than what is only novel.

14

Miscellaneous Media

To most people the term "mass media" means network television, radio, or daily newspapers. To be sure, these are the loudest voices we hear in the public arena, but there is another kind of mass media. It exists as a quieter murmuring of lesser voices, blending with the background of daily life. What is distinctive about it is that while the voices are miscellaneous in origin, there is unity in their song.

Through a hundred small ways we are, as a mass public, falsely addressed as individuals. While everyone appreciates having his individuality respected, no one enjoys having it pretended. The miscellaneous voices speak to us as if they were the utterances of one person to another. Their messages come in many forms. We meet them every day, as we do radio, television, and the press, but we do not recognize them for what they are.

Our exposure to major media is a result of a conscious

action of our own. We sit down in a chair and turn on the radio or television set or pick up a newspaper, taking time-out from our daily world. The miscellaneous media, quite differently, are a part of our real world that we encounter in the process of going about our business. They include such diverse items as junk mail, bumper stickers, pins, computerized letters, and advertising posters. It is the media of minor novelties powered by our nouveaumania.

When the toll-gate sign flashes "Thank You" after you drop a coin in the hopper, the authority that oversees the operation of the turnpike has utilized a form of miscellaneous media to communicate with you. For $29.95 you can buy a gadget to hook to the rear of your car that flashes back "You're Welcome." One makes as much sense as the other.

When you pick up your telephone and dial a number to listen to a tape recording of a prayer, a comedian telling jokes, or a voice telling you that you're the most wonderful person in the world, or a voice that will agree emphatically and sympathetically with everything you say, you are as close to communicating with the miscellaneous media as you will ever get. In July of 1974, to be told you were truly wonderful, if you were a woman, you dialed NBC Radio's Compassion Phone to hear a throaty male voice, choked with ersatz emotion, say, "Hi, y'know, I . . . I think you're the most exciting woman I've ever met. And I often wonder if the real man in your life realizes—really realizes—how lucky he is. Well, I do. And I'm so glad we could share this moment together . . ." In August there was a number for men, but if you call now, you will hear another voice, some-what firmer, telling you that the number you dialed is not in service.

When you buy an item in a store, a sales slip that says "Thank You. Come Again" is stapled across the top of the bag to seal it (so you won't steal anything on the way out and put it in the bag) by a mute, solemn clerk. This is the store management's way of sending you a "personal" message.

One frequently encounters impatience, indifference, and hostility in day-to-day transactions while at the same time being fawned over by the artificial cheer and attentiveness of the reproduced message. Image battles reality. The electric company spends millions of dollars to tell you all the wonderful things they are doing to improve your service, but if you dispute your bill, you are confronted with an intransigent bureaucracy and if you are late paying it, they will plunge you into darkness.

The message of the miscellaneous media is that we are welcome wherever we go, that our return will be welcomed, our business wanted, and our very existence deeply appreciated. Our own candid perceptions of these fostered impressions may dispute them. It does not warm the customer's heart if "Thank You" is printed on the sales slip, or a "Nice to See You Again" or "Hurry Back" sign is on the door or the wall if the clerk treats him as a nonperson. Nor is one cheered by the bumper sticker or pin that says "Have a Nice Day." A writer of those anonymous chronicles in *The New Yorker*'s "Talk of the Town" noted that this bumper sticker was the most repelling he had seen, even amid the obnoxious superpatriot slogans typical of many stickers. He wondered why this was. I think the repulsion we feel arises from our objection to the mechanization of human sentiment. We may utter this phrase mechanically to one another, but we draw the line at having a printing press say it to us, using the rear end of an automobile as a messenger.

In the age of mass communication, the citizen with no means of participating in the national conversation must resist the feeling of anonymity, of a distant view of himself among the masses. He looks for ways to talk back, hence the popularity of radio talk shows where listeners can call in; spray-can and Magic Marker graffiti in the form of names and addresses; the Letters to the Editor column; and similar places where the individual may announce his presense to the mass. In New York City subway cars, the lines of advertising posters above the seats blend in naturally with the name and number designs that are the inner city's youth's way of saying "I am." Viewed objectively, there is little difference in a hemorrhoid-medicine company plastering the car with its name and a young boy adding his with spray paint. His mark is no more a defacement than the litany of bleeding rectums, headaches, money problems, bad breath, nervous tension, and aching feet described in the ads.

Every time I open my mail I am sure to find at least one letter that I can classify as miscellaneous media because it will pretend to know me, to speak to me in a special way. For example, I recently received a solicitation to subscribe to a magazine published by the Smithsonian Institution which in a vague way suggested that being allowed to buy their magazine was an honor to me. I was notified of my "nomination" to membership as National Associate, number F270740. Would I have the courtesy of making an early reply? If I did not want to "accept" (pay for) the "membership" (magazine), they would like to "offer" (sell) it to someone else, which they are currently doing every hour on the radio.

Following my failure to respond, I have since received six more "opportunities" to become a National Associate, having apparently been forgiven for my failure to accept F270740. Later I read in the advertising section of the *New York Times* that the audience for the *Smithsonian* magazine was "suburban businessmen and their wives." Although it had sounded vaguely like a nonprofit venture, the magazine turned out to be a prime location for "sellers of autos, liquor and high class mail order items" and had sold three hundred pages of advertisements for $900,000, but "is making more money on subscriptions (memberships) than it is on advertising."[1] (The device of conferring fake honors and memberships to raise money was also used by the Finance Committee for the Re-election of the President. Contributors of $15 were given a Membership Identification Card, as "a keepsake you'll be proud of for years to come.")

A short time later I was similarly honored by an invitation to become a member of the American Museum of Natural History ("Dear Member-elect") and membership #3795925 in the Early American Society. My wife ("Dear Nominee") was offered a *numbered* membership in the National Trust for Historic Preservation." (Italics theirs.)

The American Academy of Political and Social Sciences sent me a simulated engraved invitation, with my name written in real ballpoint ink, asking me to become a member, which of course, entitled me to their magazine. The hand written invitation is also used by *Business Week* to whom I have somehow become known as a "business leader."

Everyone has his share of such offerings. There are 25,000 mailing lists currently in use containing over 5 billion names. If you are "average," your name will appear on about 150 lists. A baby in California was reported to have received over 2,000 pieces of mail by the time he was eight months old.

Because mailings that tie in membership in some organization to a subscription have been made with increasing frequency, I can safely assume they are successful. But who is deceived by this thin guise of ersatz honor? Why does it work?

The honorary-membership ploy is a small voice in the chorus of pretense by which people are, and at the same time, are not deceived, much in the same way as we enjoy a play by the willing suspension of our disbelief. Not wanting to think of ourselves as members of a mass society, we attend to anything suggesting that somehow we are known, that we have been selected from the unwashed millions.

Emma Lazarus' poem for the Statue of Liberty equated "huddled masses" with "wretched refuse," a reflection of the American ideal of uniqueness and a revulsion to being

lumped together in an undifferentiated mass. It was the
rugged individual that counted in a democratic society and
he did not see himself as lost in the crowd. Like the Marl-
boro man, he is supposed to be starkly silhouetted against
the sky, a lone figure dominating the landscape and the
distant mountains.

When people speak of the artificiality of American life
they frequently mean, I believe, the sustained pretense of
being individual and personal when it is not possible to be.
As we evolved into a mass society it became hard to maintain the image of the individual as self-reliant, innovative,
and free. As voters, consumers, viewers, and subscribers,
our number is too great and our individual influence too
slight for us to be accorded the kind of personal attention
most of us feel is our due. Mechanical novelties and techniques of mass communications are employed to hoke up
the illusion that every man is known, is unique, special, and
esteemed. Why else are we susceptible to the empty familiarity of the computer-generated letter that uses our name
not only in the salutation but throughout the body copy?
I received a letter from an insurance company that began
as if it had been translated from Russian: "We're concerned
about the risks we take every day, Truman Moore, and the
great need for protection . . ." Was I to believe that this
was a personal letter written just for me?

The *Weekly Reader*, a publication for schoolchildren, sent
out a direct-mail solicitation for a three-volume set of sex-
education books. The letter my wife and I received began,
"I'm writing you this letter because you are devoted and
concerned parents. It's a special letter about a very special
person—your daughter Rebecca" who was approaching a
"critical stage of childhood," meaning that we would need
the books to answer her questions. The letter was imprinted
with the signature of someone named Virginia Jean Noble.

There is a scene in Kafka's *The Trial* in which Joseph K.
enters the cathedral and is startled by a voice calling his
name. It is perhaps with this in mind that I find "personal-

ized" letters unpleasant. Who are these shadowy people who pretend to speak to me, to know the quality of my parental devotion, to know my daughter's needs?

John J. Pollock of Time Inc.'s Life Computer Service once said that the computer-personalized letter "reaches out to the individual and treats him like a human being." The computer, although an instrument of depersonalization, is being used to "repersonalize the world."[2] It doesn't seem possible that being called by name by a computer could give anyone a warm glow.

When Abraham Beame took over the mayor's office in New York City, he sent out a form letter asking 265 officials of the Lindsay administration to submit their resignation. While no one objected to the new mayor's right to dismiss holdover appointees, many of the departing officials strongly objected to the "impersonal tone of the letter."[3] They were like those nineteenth-century farmers who so objected to "machine-made" letters that Sears, Roebuck wrote them by hand long after the typewriter was in use in most other businesses .

Would a computer-generated letter have "repersonalized" the dismissal? Do all of us, like the departing commissioners, want our messages sprinkled with faked personal references? There were other ways the dismissal might have been carried out, but none seem to be really any better than equal treatment by form letter. The mayor dealt with a perfunctory matter in a perfunctory way. In objecting to his form of dismissal, the commissioners overlooked the fact that there were too many of them for the mayor to make much of an occasion of it. In playing the game of personalization, it is important to maintain the front that the person addressed is a discrete being, not just one of a multitude.

Isaac Asimov has found much to be hopeful of in the marriage of computer to the mailing list, since it means that the junk mail we receive will be more specifically directed to our interests. He foresees an ongoing census in which vital statistics are continuously updated for government and commercial use, allowing the advertiser to pinpoint his potential customer and address him with some familiarity. (Mistakes do occur. I know of a corporation that is frequently addressed as "Mrs." Betty Furness, listed on the Consumers Union letterhead as Betty Furness, Sec., got a letter offering her a Sec family coat of arms. Terence Cardinal Cooke received a letter in which he was addressed throughout as Mr. and Mrs. Cooke. And I was recently invited to join the Navy.)

Form-letter familiarity, even when aided by the cleverness of a computer program, is seldom convincing, but it can mislead, as did the "two-neighbor" gambit used by *Reader's Digest* in 1970-1971. A direct-mail letter informed the recipient that he and two of his neighbors, named in the letter, had been selected to participate in a $400,000 sweepstakes. The particularization of the letter led many to believe that they had actually been chosen for the final drawing, which was not the case. The same letter had been sent to 25 million people.

A growing awareness on the public's part of the degree to which such seemingly personal information can be mass-produced has reduced the effectiveness of such direct-mail techniques. Researchers involved in business surveys of top management now take great pains to keep their letters from looking like computer print-outs. To ensure a reply they have learned that "envelopes must be individually typed (not run through a computer), that the mailing piece contain no color material lest it be confused with 'junk mail,' that the letter be personalized and the signature be in blue." The survey should be "nation-wide" and for "an important organization."[4]

It was not too many years ago that direct-mail letters often began with such unspecified grossness as "Dear Sir or Madam." In an era of increased sophistication in extracting personal inferences from mailing lists, such general greetings now have the ring of honest candor, of not pretending to familiarity.

It is characteristic of the miscellaneous media that their messages are received almost involuntarily. The individual finds them everywhere, with no dial to turn or switches to flip. They are a part of our world, inseparable from the whole. In their single parts they only slightly annoy, mildly amuse, or faintly disgust. Their cumulative effect is to involve us in a larger game of pretense, quite against our will and sometimes without our knowledge.

15

The
Way
It Is

In 1927 the first telecast between cities in this country was shown in New York—an image of Secretary of Commerce Herbert Hoover reading a speech in Washington, followed by a vaudeville act from Whippany, New Jersey. The *New York Times* said the event "outruns the imagination of the wizards of prophecy."

Regular television broadcasts were made in New York, Schenectady, Philadelphia, Chicago, and San Francisco during the late thirties, but like frozen food, television did not become a part of mass culture until 1946. It made its presence felt immediately. New eating habits developed, centered around the TV set as families ate TV dinners on TV trays. Network-radio died overnight and the big mass magazines began their slide to extinction.

Television was not particularly good, but it was all new. Harriet Van Horne wrote in *Collier's* that the first year of

*149

postwar television would be "so enchanting, so wondrous in its newness, that nobody will be scornful." In later years, she said, the public would certainly not continue to buy sets just to see "old movie shorts."[1] But in the rampant nouveamania of peacetime America, people watched almost anything that flickered across the screen, from the first test pattern in the morning to the flag waving at sign-off.

People were interested in television not only as a source of entertainment, but as a subject itself. The *Reader's Guide to Periodical Literature*, during the two years between May 1945 and April 1947 listed 150 magazine articles about television, almost all of which dealt with such nuts-and-bolts topics as the technicalities of transmission and reception, industrial applications, and broadcasting hardware.

Better television generally meant a clearer picture rather than a more interesting program. The great debate raged not over what use was to be made of this new medium, but over the question of color or black-and-white, an issue the Federal Communications Commission settled in 1947 by ruling that color telecasts should be left for the more distant future even though they were then possible, though difficult.

The kind of programs that would appear on the home screen was considered a matter of fate, over which the man in the street would have no direct control. *New York Times* critic Jack Gould saw the evolution of television as technical, economic, and programatic, meaning that, by and large, the public would be shown what advertisers were willing to pay for.

Hoover had said of radio that the public would never allow so valuable a medium to be "drowned in advertising chatter." Although the medium had been drowned, the same public that had failed radio was expected to protect the integrity of television. Walter Lippmann later proclaimed its failure when he said that television had become "the servant, indeed, the prostitute, of merchandising."[2]

In no other country was control of programing left so entirely in the hands of commerce and industry as it was in

America. The interests of soap or breakfast-cereal manufacturers who sponsored the programs—and by their patron-
age, or lack of it, censored the contents—were, in effect, held
to be in harmony with the public interest, as all broadcasting
is required to be by law.

The commercialization of the mass media occurred partly
because the public welcomed first radio and then television
as novel gadgets and applauded improvements in their technical performance and ignored the social impact of the new
media.

Nouveaumania influenced the development of television,
and television has, in turn, influenced its audience. Welcomed into our homes as a kind of ultimate novelty, a
trusted and tireless new friend, it blared away to enraptured
families who sat mutely before its blue glow.

TV was hailed as the savior of domestic life because it
brought family members together, but it soon became apparent that they were not drawn closer to one another but
closer to the images on the screen. Strange new relationships
developed between the viewer and the viewed.

. Television has disrupted the audience's ability to distinguish the real from the unreal, the image of the thing from
the thing itself. Many people developed the illusion of
knowing the "personalities" they watched on television. A
schoolteacher writing recently in the *Village Voice* observed:
"The seductive thing about the media is that they let us
spend as much time every day as we wish getting to know
ghosts: celebrities, places, problems which seem so familiar
after we have seen them often enough that we forget they
really exist only in our imagination . . . We have whole
salons of imaginary friends, we are at home (as the international credit card and airlines people say) everywhere in
the televisable world, but where does it get us? . . . The
people we know through media are not our friends. And our
friends (or those who might be our friends) are not media
creatures . . . The media-believer keeps so much company
with shadows that real people, live and in color, don't reg-

ister on his senses unless they are standing dead in front of him talking loud."[3]

People and places seen regularly on television become intertwined with our real-life experiences. After the President's assassination I had to make several business trips to Dallas. On the last trip, as I looked at the Texas Book Depository from the expressway, I was unable to remember if I had seen that view on television or on my previous visits. I was startled to realize that the images had run together and reality had mixed with half-remembered TV coverage.

Sometime later I was reminded of this when a friend of mine repeated a remark he had heard on Johnny Carson's show. Instead of saying "I heard Johnny Carson say . . ." my friend began "Johnny told me . . . " He corrected himself with some embarrassment.

The individual behind such a televised image has the special quality of being able to "achieve an intimacy with what are literally crowds of strangers," an intimacy which is "an imitation and a shadow of what is ordinarily meant by that word."[4]

Since the projected image is never allowed to change and may be terminated by the turn of the channel knob, the nature of the phantom relationship is nonenvolving and undemanding, easily broken, and devoid of development. You can watch Johnny Carson for a year and gain no insights into his character beyond what little could be surmised from a single performance. What is presented to the camera is a highly selective and carefully tailored, consistent Personality.

There is a loosening of the social fabric as the direct interpersonal bonds are broken and attention is transferred to immature relationships with Personalities, tending to isolate and alienate the individual from real people and create in society what in physics is called an unstable mass.

Just as we substitute plastic imitations for the articles of our culture, we substitute Personalities for friends, talk shows for conversation, and programing for reality. Television, because we spend so many hours in front of it, has become one of the major suppliers of our visual images and ideas about the world. Family life, under the influence of television, has become a situation comedy in which Dad is a bumbling but well-meaning fool and Mom a long suffering font of ironic wisdom who divides her time between controlling the destinies of those around her and admiring her waxed floor or sparkling sink.

Doctors are invariably competent, private investigators cleverer than the police, and FBI agents incapable of error or misjudgment. And by constant repetition of the television commercial, a new set of ethnic stereotypes has been created in which the Italians overeat, the Jews get upset stomachs, the WASPS get headaches, and the Poles in hard hats get sandwiches in plastic bags.[5]

The most dedicated viewers are children, who, according to a CBS documentary, watch 25,000 commercials a year, or 220 minutes of television ads a week, and spend more time watching the programs than they do listening to their teachers, recalling David Riesman's description of the American child as "consumer-trainee."

Children's programs have drawn criticism for the amount of violence they depict, but many researchers believe that it is the passive activity of watching television that affects children as much as what they watch.

Dr. Douglas Heath, a professor of psychology at Haverford College, said that in his travels around the country he found schoolteachers complaining that children get bored in school if they are not constantly entertained. They are losing their own imaginative potential—they don't even know how to play. "*Sesame Street* gives children their first thousand hours of education. It's terribly fast-paced. When a five-year-old gets to school he expects the same performance from the teacher." Heath feels that such programs may to a limited degree accelerate the child's learning (which

Piaget calls "the American obsession") by teaching him a smattering of various subjects, but it doesn't teach him how to learn, and in fact interferes with the process.

The television format encourages the "frenetic search for novelty and excitement" which works against the development of personal stability. Too much novelty, he says, induces dilettantism when it is persistence that children need to learn.[6]

A marine biologist I talked to suggested that the continual excitement and drama of such highly regarded programs as Jacques Costeau's undersea epics mislead children. The weekly succession of end results conceal the drudgery, the frustration, and the infinite patience required in all sciences. It is difficult for young people to realize how much routine hard work and long preparation are necessary when everything looks quick and easy.

In *The Disney Version*, Richard Schickel wrote that the function of art is to expand consciousness and the function of the media is to reduce it. When program content is designed for large audiences, a kind of reduction occurs. A television writer told a Senate hearing, "They allow laughter but not tears, fantasy but not reality, escapism but not truth . . . Why cannot television ever reflect the real world? Why must it always be that television brings you a detective in a wheelchair and that is a success so next they bring you an insurance investigator who is blind and that is a success with the result that ABC is trying to put together a new show about—I wouldn't kid about this—a sheriff in the Old West with a stiff trigger finger."[7]

Faced as we are with the banality of most programing it is worth remembering that the promise of television was that it would open a wider world to its viewers. When television was "wondrous in its newness" all the performing arts were to be made accessible to the citizens of the remotest village. The national level of culture would be raised immeasurably by widespread broadcasting of concerts, ballets, plays, and operas. It was believed that television would revive the theater, particularly aiding local groups, since the demand for performances by television stations all over the country would provide them with money and audiences.

This optimistic outlook prevalent in the postwar era placed innocent trust in the new broadcasting technology. There were formidable dissenters, however. James Petrillo, head of the musician's union, refused to permit musicians to perform on television until it was clear just what effect the new medium would have on their livelihood.

While more people have watched the performing arts on television than have ever seen them live onstage, it cannot be said that television has increased the individual's appreciation of them. There is a vast difference between ballet seen on television and seen in person. On the small screen it has none of the human presence essential for its aesthetic impact. The millions of people who watch an occasional ballet on television are given the false sense of having seen a real performance instead of an image which contains none of the vitality to make it a compelling experience. Television's supposed promotion of the performing arts may eventually do them more harm than good.

More is not better. The mass production of television images does not necessarily broaden our understanding of the world. This is particularly applicable to news programs where the focus on events is reduced and simplified by the television format. The result is more and more about less and less. Television reporters fight for more time, and television executives want more ten-second stories. Writers want a deeper exploration into the human condition, and sponsors want to appeal to millions and offend none. What comes out of this conflict is the highly structured news or entertainment program which helps form, over time, our narrowed view. Reuven Frank, former president of NBC News, said, "There are events which exist in the American mind and recollection primarily because they were reported on regular television news programs."[8] The opposite is equally true. There are events which do not exist because they were not reported, or they exist the way they do because of how they were reported.

Edward Jay Epstein, in an analysis of television news reporting, gave a number of examples of stories that were not reported because of network policy—the riots after the death of Martin Luther King, the Vietnam fighting during a time when it was decided that the peace negotiations were more important. Television news has become the chief arbiter of reality for many millions of people. It is not, as Epstein clearly shows, a mirror held up to reflect society. Mirrors do not select, omit, rehearse, transform, or dramatize as television news reports do on a routine basis."[9]

Television, by its very nature, has an influence on what it reports. A large portion of the news is tailored to fit the limitations of video coverage with its cumbersome equipment. In covering events in the city as a photographer, I always found that lights, large video cameras, tripods, microphones, and electrical cables marked the spot where news was shortly to occur on such formal occasions as press conferences, meetings, and speeches. The need to have crews available and on location in time to set up restricts television news coverage to scheduled demonstrations, parades, and airport arrivals. The need for coherent film footage revived the picket line and the placard and made them new weapons of protest, confined though they might be behind gray police barricades.

The presence of a television crew even creates its own news. Participants are in some degree stimulated by the knowledge that the whole world is watching. I witnessed a graphic example of this television effect from the marquee of a Times Square Hotel, waiting for the arrival of midnight on New Year's Eve. Shortly before the hour, the television lights were turned on the crowd, which was packed solidly for several blocks. Within the circle of light, people cavorted and cheered, creating a wavelike pattern of jostled bodies. Just beyond the edge of the spotlight—out in the untelevised darkness—the crowd was silent and calm, watching, enviously I thought, the bright dancers. The light divided the crowd into spectators and performers as it swept Times Square. (Someone once made the comparison between the objects of televised news and light sensitive chemical elements which could never be seen, since they were physically altered by the light that rendered them visible.)

In more recent times, the demands of television for "newer" news has begun to exert a more direct influence over the events that are covered. For example, the FBI shoot-out with the SLA (itself a media creature) was delayed for two hours so that technicians could prepare for live color transmission.

Part of television's distortion is caused by our nouveaumania, which demands that news be not just a recitation of the day's events but continually different as well. Long-running stories, such as the Watergate affair or the CIA investigation, must be constantly given a new "angle" lest the public lose interest. One of the chief strategies of the Nixon White House was to practice informational overkill, hoping to drown our interest in a sea of transcripts and press releases.

Television has the impact to swiftly focus public opinion on a new cause and the motivation to do so because of its own needs to provide novelty in news programs and documentaries. However, attention gained in this way is hard to hold for long. Ecology, once a fresh subject visually suited for television, was worn out long before the rivers were cleaned or the air made pure. De Tocqueville noticed the tendency of Americans to rush from one subject to another on the assumption that their intense public interest had, by itself, solved whatever problem had attracted it.

The pressure of nouveaumania on news affects the behavior of those who compete for its attention. It has a tendency, for example, to radicalize social movements, and the civil rights and women's liberation movements may serve as examples. In both, once the public had digested the statements of the more reasonable or moderate spokesmen, the increasingly radical voices captured media attention until it seemed that a competition developed to see who could find the most extreme language to present the latest, hence the most newsworthy, state of the movement. To a large extent, both movements burned themselves out on the incendiary rhetoric of their media-appointed leaders.

In a somewhat similar way, novel behavior is usually reported as a part of a new movement. A man running naked down Main Street may be reported as news because the subject of streaking is a news fad. When it is no longer a novelty, a man displaying himself in Kansas will not be reported in New York. Local events are reported as if they were related to events in other parts of the country, or reflections of the mood of the country. In Epstein's words. "Because network news, essentially, has to be national news, network producers . . . tend to view events not as isolated incidents but as threads running through the entire fabric of the society . . .Through this prism a student strike at a single college is seen as 'symptomatic' . . . of student discontent throughout the nation . . .[10]

In recent years news-media reports depict campuses as quiet. We no longer see student disputes with the administration either because they are not happening or because they are not news. Since we must rely on news reports for information about the rest of the country, it is difficult to know what is happening when the media are not looking.

It is largely through television that we are made conscious of the world at this moment, as it is now. Newspaper reporters broke the Watergate scandal, but it did not penetrate national consciousness until it made the news on television.

Every night at the end of his program, Walter Cronkite says, "That's the way it is," and millions of viewers take him literally. His more accurate parting comment might be that in the minute sector of the vast range of human events which the editors and the network officials had arbitrarily chosen to be shown in a brief, selective, and highly structured way, that's the way it has been made to appear. After twenty-three minutes of news, interrupted for seven minutes of commercials, it is easy to believe that one is informed about the world, just as it is easy for the newscaster to believe that he has informed us.

16

The
Artificial
American

Up until the Industrial Revolution, the world did not change very much in a lifetime, but a man today looking back on his youth feels as if he were witness to ancient history. Movies like *American Graffiti* present even the early fifties as a period piece and disc jockeys play turn-of-the-decade music as golden oldies.

A lifetime, as a measure of human progress, can dramatize the swiftness of change. Austin Moore, my great-uncle, was born in 1846 and died in 1946. At the time of Uncle Austin's birth, James K. Polk, the eleventh President of the United States, was in the White House and the population of the country was around 20 million. Of the ninety inventions considered the most influential to modern life, sixty occurred in his lifetime.

Scarcely less impressive to me as a boy was my father's birth date of 1899. It amazed me that he predated those in-

dispensable elements of civilized life, the radio, the telephone, the automobile, and the airplane. I enjoyed num-
erous accounts of his prehistoric youth. "In those days we
didn't have such-and-such like they do now." I could never
understand how people managed without the modern
conveniences and I felt superior to those unenlightened
citizens of the past to whom my everyday experiences would
have seemed miraculous.

My daughter, at eight, was equally amazed that I, at her
age, had never seen television or a jet plane. She, who has
been to Europe and the Caribbean, could not believe that
as a small child I had never ventured beyond a few Southern states.

In no other country and in no other time have people
grown up with such rapid change a fact of life as they have
in this country in this century. In my own experience, I
grew up with a constant awareness that the world was
changing, that change was for the better, every day and
in every way.

The differences in life style between my great-uncle's life
and my father's life, what my life has been and what my
danghter's adult life will be, can be measured in large by
the way things are done.

A new way of doing something (shopping, driving, building a house) has been more easily accepted in America than
in Europe. We are less traditional here because we have
never seen the value of the past, nor have we, until recently,
begun to realize that our selves and our society are shaped
by the daily experience of accomplishing routine tasks. A
man who tends to a horse is different from a man who
tends to an automobile or an airplane, and when we change
from horses to automobiles and airplanes, we change more
than mode of transportation.

Before I had a bicycle, my exploration of the world was
limited to what I could reach on foot. I was more aware of
the minute details of my neighborhood. With a bicycle I
could venture farther, and the attention I paid to the nuance

of my surroundings diminished a little, for I measured distance in miles instead of blocks.

I got my driver's license when I was fourteen years old, and the first time I drove a car I was aware that more was changing than gears. The relationships between me and my parents, and my friends, and the world changed. I had gained a measure of independence from the limitation of time and space.

At about the same time, my cousin Bob bought a bi-wing open-cockpit airplane and took me flying and I saw that towns were but clearings in the forest and were connected by nearly straight lines. The beach, which had seemed to be a universe of its own, was a narrow ribbon of sand curving into the distant haze. The view from the plane demystified and simplified the world, and my newly gained mobility brought it all within my easy reach.

I felt somehow less tied to my particular clearing in the woods or spot on that white ribbon of sand. There were many other spots and the ribbon was long. All, though distant, were accessible. The airplane thus enlarged my view of the world and at the same time impoverished it.

A little later Dwight Lambe, my summer employer, took me with him on a crosscountry flight in a Cessna 140. We left Myrtle Beach early in the morning and stopped in Savannah and ate breakfast at the airport. We flew on to Jacksonville and had lunch, arriving back home in time for dinner. I ate each meal that day in a different state. Walls fell in my mind and spaces of green and blue opened up beyond the place where my imagination of the world had always stopped. Savannah, as distant to me as Singapore, was now only a short hop. Both Jacksonville and Savannah, which had been far-off places, now existed as airports and coffee shops, nearly identical.

The transition from foot to bicycle to car to airplane, which I experienced in the course of a little over two years, changed my whole outlook. It set off a train of thought that led to notions about the possibilities of life, the imaginary nature of limitations, and a sense of familiarity with a wider world.

Social change occurs when many people more or less simultaneously adopt new ways of doing things and the accumulation of new ways changes their outlook. This may be good or bad, depending on viewpoint and circumstance, but the visible change in any of the components of our daily life also has invisible results, both immediate and long-term. The supermarket was a change from the store on the corner. The housing development was a change from individually built neighborhood houses. The automobile was a change from the horse and buggy. The importance of these changes cannot be seen by looking at stores or houses or cars but by looking at the way life has been changed by them, gradually, until the familiar day-to-day world becomes a strange place in comparison, although perception of the change may come suddenly.

I have a clear memory of a Saturday morning during my sophomore year in college of suddenly sensing a new way of life that had developed since I acquired my fixed notions of society. I was picking up my laundry at a shopping-center laundromat and as if seeing for the first time—to use a popular movie line—I looked at the rows of development houses that covered the rolling hills and caught wind of a dismal life. And there was something impersonal about the vastness of the supermarket and the glaring brightness of the fluorescent lights that did not seem to bring out the best in people. The women seemed tired and frayed, the children fretted, the men looked bored, as if enduring life wearily. With shopping carts laden with large-economy-size boxes of soap flakes and cereal, they trudged homeward across the huge parking lot.

It has been said that the suburbs attracted mainly people who wanted to be with their own "kind," hence the underlying drive to move was a search for identity. It is ironic that what they found was a souless dislocation—a loss of identity. A generation of postwar novelists, reporters, and sociologists have sifted the ashes of American family life burned out in the urban sprawl and the suburban wasteland. None of the builders and planners had any idea of the consequences of the combination to which they contributed, and none of the victims realized the hazards to which they so eagerly ran.

Nouveaumania is not only an addiction to novelty but an insensitivity to the results of change deliberately instituted as in the case of the housing-development-shopping-center complexes. In our readiness for the latest thing, we often overlook the value of the lasting thing, and underestimate the impact of the new. Communication satellites offer a timely example. Dr. Robert Jastrow, director of NASA's Goddard Institute for Space Studies, has called them the fifth revolution in human communications after speech, writing, printing, and telephone and radio. Five domestic communication satellites will be in operation by 1976 and "the fifth revolution will have taken place [with] a quantum jump in the capacity of Americans to talk to and see one another."[1]

Dr. Jastrow predicts that some of the short-term effects of this new network will be that in eight years a telephone call halfway around the world will cost as little as a call next door. A single world language may develop, probably English, and because of the ease of communications, people will not need to live in cities. Dr. Jastrow quotes Arthur C. Clarke, whom he sees as "the most accurate forecaster of the future around." Clarke sees "a slow but irresistible dispersion and decentralization of mankind . . . Megalopolis may soon go the way of the dinosaur."[2]

In the long run, what Dr. Jastrow suggests is far more important. He sees a parallel between instant global communications and multicellular communications. The satellites will supply mankind with a central nervous system, creating a unified global society, a change comparable to that which occurred "several billion years ago when multicellular animals evolved out of colonies of individuals." This will create political stability and order, but at the expense of individual freedom. "Just as a cell cannot migrate from, say, the brain to the liver, the freedom of the individual will be restricted in the new society." But not to worry. Those living now will be dead by then and our descendants will be "conditioned to accept their more restricted options."[3]

Our ability to have some control over the direction of our lives depends on whether or not we are able to look at communication satellites as cheaper telephone calls or encroachments on personal freedom—short-term or long-term. Unless we are to be tossed about by succeeding waves of change we will have to learn to look at those innovations we initiate in a more imaginative way.

Always in the past we have tried to do whatever technology made possible. From the perspective of a few years, the decision not to build a supersonic airliner appears to mark a change in American thinking. We were able *not* to build a plane just because it was technically more advanced.

An awareness is developing in this country that constant change is not necessarily human advancement, nor is it always a sign of health, either in nations or in people. In fact, one of the most prevalent mental illnesses—depression, the common cold of psychic disorders—may be traced to our nouveaumania. Eight million people in America are said to be depressed enough to warrant medical attention. Some researchers believe that the causes may be a "symptom of the culture" prompted by too much change in values, too much rootlessness and moving around.[4]

In *The Pursuit of Loneliness*, Philip Slater defined "the old culture," in this case the predominant one, as a culture that "worships innovation and turns its back on the past, on familial and community ties." In the revolutionary sixties he found a counterculture, "the new culture" which was "preoccupied with tradition, with community, with relationships—with many things that would reinstate the validity of accumulated wisdom. Social change is replete with paradox, and one of the most striking is the fact that the old culture worships novelty, while the new would resuscitate a more tradition-oriented way of life."[5]

The new culture remains embryonic, its growth uncertain. The rapidity with which we discard our possessions and bulldoze our surroundings is some indication of our continued willingness to accept the impermanence of the old culture as a way of life. Bizarre examples abound. In California you can buy an instant home—a condominium completely furnished with paintings (choice of ocean and sailboat scenes, rocks and streams, or hard-edged) and furniture (choice of four basic styles—chrome-and-glass, farmhouse contemporary, a combination of the two, or a "design in walnut and oak"), down to the stainless-steel flatware, flower arrangement (one), sheets and towels, and ashtrays. A computer keeps track of who orders what to avoid having identical apartments next to each other. One man who purchased one of these ready-made homes was asked if he minded living with someone else's taste. "Lots of people have the same kinds of cars and they think nothing of it,"[6] he replied.

When we reach out for tomorrow with both hands, we necessarily turn loose everything else. Implied in our nouveaumania is a disregard for the old (i.e., outdated), and apathy toward maintaining what we have, if not outright hostility. Remnants and reminders of the past struggle for survival. In New York City there are 1,100 statues, plaques, tablets, and other monuments, but most have been damaged because "many New Yorkers take glee in writing on, smashing, toppling, sawing, painting, shooting, stoning, stealing and otherwise defacing monuments."[7] The destruction of the city by bulldozer and wrecking ball has been no less intense. New York seems to grow by digesting its past—a poorer diet with each passing year.

That we have little interest in the "beautification" of our cities and countryside seems natural in the light of our greater interest in tomorrow. ("America, of thee I sing; Sweet land of Burger King," wrote Ada Louise Huxtable, on the tearing down of a historic 116-year-old house to make way for a hamburger stand.)[8]

Charleston, South Carolina, in spite of the Civil War destruction, has preserved a large part of its architectural past. A citizens' group in 1974 barely prevented a developer from Atlanta (an industrial city to the north) from knocking down a block of ante-bellum buildings and putting up an eight-story condominium apartment house. Charleston, like many Southern cities, has trailed behind the rest of the country in industrial progress and growth, but as the sterile, alienated Shoebox Modern buildings impose a dulling monotony to the appearance of other American cities, Charlestonian backwardness is not without virtue.

One need only travel through our country to find out that we don't greatly value its appearance, although there are occasional signs of hope where one would least expect them. When Los Angeles county put $75,000 worth of plastic bushes along a boulevard, local citizens removed them in protest.

The most prominent features of the American landscape bear witness not to our sense of beauty, but to our devotion to the automobile, electric power, and neon. What one sees most are gasoline stations, drive-ins, highways, power transmission lines, and huge signs. A visitor from outer space would have difficulty deciding whether the planet was inhabited by cars serving people, or people serving cars. The mobile-home trend seems to indicate a marriage between novelty and mobility in which the American is formally wedded to his automobile and is thus made heir to the industrial tradition. According to the U.S. Census, there are 1,850,000 year-round mobile homes in America, and in 1970 they accounted for half the new houses sold. High mortgage rates undoubtedly accounted for much of this, but it also represents a real preference. One house-trailer resident said, "What I like about a trailer is that I can trade it in every three years and get me something new."[9] And a Vermont couple who got married in a parking lot in their camper used almost identical language. Said the bride, "We just wanted to have something new, something a little different."[10]

Professor J. Wreford Watson of the University of Edinburgh, explaining America to his fellow Britains, said, "The geography of newness has become an essential part of America. Newness is adopted almost for its own sake. The passion for newness dominates the American view. Innovations and discards, expressways and scrap heaps go side by side . . . the landscape is littered with the relics of outdated roads and bridges, with torn-up railways, with abandoned farms, with dying villages, and with ghost towns—with all, in fact, that has been superseded by the new and better."[11]

A recent addition to this geography is the new "towns" (perhaps future cities or ghost towns) that are rising from the 13,500 intersections of the Interstate Highway System. At these junctions motorists stop for "gasoline, souveniers, camp ground spaces, and 'radar' range hamburgers." While the resident population of these nameless crossroads may number less than a dozen, a transient population of many thousands is "lured off the road by the garish gasoline beacons . . . to convenient, if rarely attractive, places for rumpled travelers to stretch their legs, fill their tanks, walk their pets, quench their thirst, satisfy their hunger, relieve themselves and be on their way."[12]

Somewhere in the thousands of pages of Thomas Wolfe lies a passage about a train crossing America, felt a long way off through a hand touching the rail, sensing the approach through the distant vibrations of the locomotive's enormous, speeding weight. Wolfe imagines it crossing the wheat fields, struggling through the mountains, its searchlights knifing through the darkness as the train races across the continent. It is a lyrical description of the great beauty Wolfe found in the American landscape.

The train that approaches New York City from the South runs through a shallow, swampy tract of land, grayed by smoke from garbage dumps, petroleum refineries, and corroded chemical works, all emitting a tangle of eye-watering odors. Perched atop fetid heaps of rubbish is a line of six billboards raised on tarred poles to the eye level of passengers on the train. Here amid the industrial carrion, viewed through cracked and dirty train windows, are advertisements for restaurants and Broadway musicals, monuments of the "geography of the new" and its overwhelming inappropriateness.

Malaise and junkyards, street crime and litter, waste and corruption, cynicism and depression represent one side of America, opposed by optimism and courage, basic decency and fair-mindedness, curiosity and inventiveness. A foreign journalist visiting America wrote that to see the past one went to other countries, but he came to America to see "a projection of life in his own country . . . five, ten, twenty years from now."[13] One wonders what he saw here of his own future. Which America did he go home with?

We yearn for those positive values, to see the great beauty, to make the country live up to its promise. We are stuffed on the rich cake of consumerism and yearn for hearty bread, tired of hearing that we are the richest and most powerful nation on earth when we are apparently too poor and ineffectual to provide ourselves with a clean and safe place to live.

We have been distracted from the business of human betterment by our love affair with excessive novelty, but the romance shows signs of abatement. What Uncle Austin saw as a parade of progress looks more like a merry-go-round today, more of the same going in a circle. It's time not to stop it, but to straighten it out and make it go forward again.

We have passed many things in our brief history and shall not see them again. It is symbolic of the lost past that even the eagle is threatened with extinction. It is symbolic of the future that his survival now depends, according to recent reports, on artificial insemination.

Among the new products that have been prepared for the

Bicentennial is patent number 3,645,026—a plastic flag that can be rapidly assembled, described by its inventor as the "world's first indoor flag."[14] Will these plastic banners, fluttering in the breezes of air conditioners, as artificially bred eagles circle overhead, be the appropriate symbols for America entering her third century of existence?

Hopefully we will discover that, in the words of Archibald MacLeish, the "greatness of the American past, the unused and wasted greatness, which government neglects, and the intellectuals, out of ignorance, despise, and the young, inspired by both, ignore, is the fact unexampled in history, that the American past contains the American future, or would contain it if we . . . could trust it again."[15]

Notes

I THE POSTWAR PROMISE

1. Francis Moore, *Charlotte Observer* (October 27, 1974).
2. Angela Taylor, "Twenty-five Years Ago Levittown was a Joke, but Today It's Thriving," *New York Times* (April 18, 1972).

II THE CULTURAL TRANSPLANT

1. Henry Bamford Parkes, *The American Experience.* New York, Vintage Books, 1947.
2. Earl Warren, *A Republic, If You Can Keep It.* New York, Quadrangle, 1972.
3. Alexis de Tocqueville, *Democracy in America*, Vol II. New York, Vintage Books, 1954.
4. "Notes on People," *New York Times* (September 7, 1972).
5. Chen Yuanchi, "And Now a Chinese Description of the U.S." *New York Times* (February 23, 1972).

III BUYING HAPPINESS

1. Paul M. Mazur, "Mass Production—Has It Committed Suicide?," *Review of Reviews* (May 1928).

*173

IV NEW!

1. Robert Kleinman, *U.S. News & World Report* (October 24, 1960).
2. Russell Baker, "Tired of That Old You?," *New York Times* (June 6, 1972).
3. *Business Week* (March 4, 1972).
4. Marylin Bender, *New York Times* (June 11, 1972).
5. Robert Townsend, *New York Times Book Review* (April 30, 1972).
6. *Business Week* (March 4, 1972).

V IT'S SWEEPING THE COUNTRY

1. Leonard Sloan, *New York Times* (December 17, 1972).
2. *Changing Times* (May 1973).
3. "From the Genius of Rembrandt," an advertisement for the Franklin Mint, *New York Times Magazine* (February 18, 1973).

VI THE COMPULSION TO IMITATE

1. Anne Hollander, "The 'Gatsby Look' and Other Costume Movie Blunders," *New York Magazine* (May 27, 1974).
2. New York *Post* (March 2, 1972).
3. J. C. Furnas, *The Americans*. Putnam, 1969.
4. *Ibid.*
5. Anne Hollander, "Art Deco's Back and New York's Got It," *New York Magazine* (November 11, 1974).
6. Keith Sward, *The Legend of Henry Ford*. New York: Rinehart, 1948.
7. Ada Louise Huxtable, "Pow! It's Good-by History, Hello Hamburger," *New York Times*, March 21, 1971.

VII THE GENERATION AS MODEL YEAR

1. Bennett Berger, "How Long Is a Generation?," *British Journal of Sociology* (March 1960).
2. David Riesman cited, *ibid.*
3. James S. Kunen, "The Rebels of '70," *New York Times Magazine* (October 28, 1973).
4. Steven Gordon Crist, undated article in airlines magazine (ca. 1973).
5. *Villager* (September 27, 1973).

6. Iver Peterson, "Race for Grades Revives Among College Students," *New York Times* (November 21, 1974).
7. Richard Lingeman, "There Was Another Fifties," *New York Times Magazine* (June 17, 1973).
8. Bennett Berger. 'Teen-Agers Are An American Invention," *New York Times Magazine* (June 13, 1965).

VIII   FASHION AS NEW SELF

1. Diana Vreeland, *The '10's, the '20's, the '30's: Inventive Clothes, 1909–1939.* The Metropolitan Museum of Art, 1974.
2. Blair Sabol and Lucian Truscott IV, "The Politics of the Costume," *Esquire* (May 1971).
3. Amy Gross, "Notes on Turning 30: Goodbye Groovy, Goodbye Wow," *Mademoiselle* (September 1972).
4. *Vogue* (January 1973).
5. Bennett Berger, "How Long is a Generation?," *British Journal of Sociology* (March 1960).
6. Gloria Emerson, *New York Times* (October 5, 1973).

IX   WOMANKIND

1. *Scholastic* (March 5, 1945).
2. Margaret Barnard Pickel, "How Come No Jobs for Women?," *New York Times Magazine* (January 27, 1946).
3. Jo Freeman, "The Origins of the Women's Liberation Movement," *American Journal of Sociology* (January 1973).
4. Molly Haskell, "Can a Woman Have a Good Marriage, Children, a Satisfying Career, a Social Life and a Super Sex Life, All at the Same Time?," *Mademoiselle* (June 1973).
5. Diana Trilling, "American Scholar Forum," *American Scholar* (Winter 1972).
6. *New York Times* (April 1, 1972).
7. Elizabeth Hardwick, 'Suicide and Women," *Mademoiselle* (December 1972).
8. Elizabeth Janeway, "American Scholar Forum," *American Scholar* (Winter 1972).
9. *Yale Law Journal* (April 1971).
10. Professor Freund cited in Nick Thimmesch, "The Secual Equality Amendement," *New York Times Magazine* (June 24, 1973).
11. Michael Korda, "What Men Won't Say to Women about Women's Liberation," *Glamour* (September 1972).

12. Catherin Breslin, "Waking Up from the Dream of Women's Lib," *New York Magazine* (February 26, 1973).
13. Betty Friedan, "We Don't Have to be That Independent," *McCall's* (January 1973).

X THOROUGHLY MODERN MARRIAGE

1. Dr. Carle Zimmerman, quoted in *Life* magazine (March 24, 1947).
2. *Time* (February 17, 1947).
3. Della Cyrus, "What's Wrong with the Family?," *Atlantic Monthly* (November 1946).
4. "How to Save the Family," *Time* (November 10, 1947).
5. Eric A. Johnson, "Everybody's Problem—Everybody's Interest," *Survey* (June 1948).
6. C. E. Ramsey and L. Nelson, "Changes in Values and Attitudes Toward the Family," *American Sociological Review* (October 1956).
7. Robert Ardrey, *The Social Contract*. New York, Dell, 1970.
8. Daphne Davis, "Moving In: The Logistics of Living Together," *New York Magazine* (February 19, 1973).
9. Lanie L. Jones, "Married or Not, Couples Living Together Have Similar Problems," *New York Times* (October 14, 1973).
10. Cited in Gail Sheehy, "Can Couples Survive?," *New York Magazine* (February 19, 1973).
11. Martha Weiman Lear, "Save the Spouses, Rather than the Marriage," *New York Times Magazine* (August 13, 1972).
12. Natalie Gittelson, *The Erotic Life of the American Wife*. New York, Delacorte Press, 1972.
13. Peter L. Berger and Hanfried Kellner, "Marriage and the Construction of Reality," in Hans Peter Dreitzel (ed.), *Recent Sociology, No. 2*. New York, Macmillan, 1970.
14. Michael Korda, "If You Look at Marriage as an Instrument of Personal Growth, Will You Be Disappointed or Not?," *Glamour* (March 1972).
15. Anthony Storr, *The Dynamics of Creation*. New York, Atheneum, 1972.
16. Cited in Lear, *op cit.*
17. Lawrence Fuchs, *Family Matters*. New York, Random House, 1972.
18. *New York Times* (May 3, 1974).

19. Cited in Erica Abeel, "Divorce Fever: Is It an Epidemic?," *New York Magazine* (November 4, 1974).
20. Fuchs, *op cit.*
21. Philip Slater, *The Pursuit of Loneliness*. Boston, Beacon Press, 1970.

XI   AUTOMATED CHILDHOOD

1. New York *Post* (October 26, 1973).
2. Lin Yutang, *The Importance of Living*. New York, John Day, 1937.
3. Barbara Seaman, "Dear Injurious Physician," *New York Times* (December 2, 1972).
4. Ernest W. Page, Claude A. Villee, and Dorothy B. Villee, *Human Reproduction*. Philadelphia, Saunders, 1972.
5. *Ibid.*
6. Cited in Ester Conway and Yvonne Brackbill, "The Effects of Medication of Fetus and Infant." Monograph of the Society for Research in Child Development, Series 137, Vol. 35, No. 4. Chicago, University of Chicago Press, 1970.
7. Dr. T. Berry Brazelton, *Redbook* (February 1971).
8. Waldo L. Fielding, M.D., and Lois Benjamin, *The Childbirth Challenge: Commonsense versus "Natural" Methods*. New York, Viking, 1962.
9. Irving I. Lewis, "Government Investment in Health Care," *Scientific American* (April 1971).
10. C. H. Peckham and R. W. King, "A Study of Intercurrent Conditions Observed During Pregnancy," *American Journal of Obstetrics and Gynecology*, Vol. 87 (1963).
11. Dr. Watson A. Bowes, Jr., "Obstetrical Medication and Infant Outcome: A Review of the Literature." Monograph of the Society for Research in Child Development, Series 137, Vol. 35, No. 4. Chicago, U. of Chicago Press, 1970.
12. Conway and Brackbill, *op. cit.*
13. Bowes, *op. cit.*
14. Alvin Toffler, *Future Shock*. New York, Random House, 1970.
15. Margaret Mead, "Working Mothers and Their Children," *Manpower* (June 1970).
16. Martin I. Heinstein, "Behavioral Correlates of Breast-Bottle Regimes . . ." Society for Research in Child Development, Vol. 28, No. 4. Chicago, U. of Chicago Press, 1963.

17. Bernadine Morris, *New York Times* (January 4, 1974).
18. Werner M. Mendel, "Other Notes on the Kibbutz," *Voices— The Art and Science of Psychotherapy*, Vol. 6 (1970).
19. John Bowlby, *Attachment and Loss*. New York, Basic Books, 1969.
20. Mead, *op. cit.*
21. William V. Shannon, "A Radical, Direct, Simple, Utopian Alternative to Day-Care Centers," *New York Times Magazine* (April 30, 1972).
22. "Cost and Quality Issue for Operators," from *A Study in Child Care*, sponsored by the Office for Economic Opportunity. U.S. Department of Health, Education and Welfare, 1970.
23. Cited in Richard Flaste, "Child-Rearing: A Return to Discipline, Without Forgetting Love," *New York Times* (December 3, 1973).
24. Fuchs, *op. cit.*
25. "The Latest Fashion: Kids," *New York Times* (June 29, 1972).
26. *Ibid.*
27. Betty Rollin, "Motherhood: Who Needs It?," *The Future of the Family*, edited by Louise Kapp Howe. New York, Simon & Schuster, 1972.

XII   THE FRENCH-FRIED CONNECTION

1. Jane Holt, "New Food Products," *New York Times Magazine* (October 21, 1945).
2. E. J. Kahn, Jr., "The Coming of the Big Freeze," *New Yorker* (September 14, 1946).
3. Win McCall, "Things You'll Eat," *Better Homes and Gardens* (April 1946).
4. G. O. Kermode, "Food Additives," *Scientific American* (March 1972).
5. McCall, *op. cit.*
6. Mimi Sheraton, "The Burger That's Eating New York," *New York Magazine* (August 19, 1974).
7. *Consumer Reports* (November 1973).
8. *New York Times* (February 25, 1971).
9. *New York Post* (September 26, 1974).
10. *New York Times* (January 18, 1973).
11. *Ibid.* (July 30, 1974).
12. John Hess, *ibid.* (July 19, 1973).

13. John Hess, *ibid.* (January 3, 1974).
14. Caroline Bird, "Gearing Up for Near Zero Growth," *Signature* (March 1973).
15. *Canner and Packer* (June 1973).
16. Cited in "F.D.A. Chief Sees Rise in Prepared Food Use," *New York Times* (March 30, 1974).
17. Gael Greene, "The Kitchen as Erogenous Zone," *New York Magazine* (September 25, 1972).
18. Angela Taylor, "The Problem: To Do a Yearbook That's Actually Something Else," *New York Times* (May 17, 1973).

XIII TERMINAL NOUVEAUMANIA

1. *New York Times* (January 3, 1974).
2. *Ibid.* (March 13, 1972).
3. *Advertising Age* (June 4, 1973).
4. Richard D. Lyons, "A Preservative Linked to Disease at Senate Hearings on Additives," *New York Times* (September 22, 1972).
5. Ben F. Feingold, *Why Your Child Is Hyperactive.* New York, Random House, 1975.
6. Lucinda Franks, "F.D.A., in Shift, Tests Pediatrician's Diet for Hyperactive Children," *New York Times* (February 5, 1975).
7. *New York Times* (September 23, 1974).
8. "'Alternate Foods' in School Lunches Opposed by Group of Nutritionists," *New York Times* (May 15, 1973).
9. William Robbins, "Nutrition Study Finds U.S. Lacks a Goal," *New York Times* (June 22, 1974).
10. *Consumer Reports* (October 1974).
11. *New York Times* (October 22, 1970).
12. Adelle Davis, *Let's Get Well.* New York, Harcourt, 1965.
13. "An Open Telegram to the Food and Drug Administration," *New York Times* (October 19, 1970).
14. Harold M. Schmeck, Jr., "F.D.A. is Urged to Curb Sugar in Breakfast Cereal," *New York Times* (August 2, 1974).
15. *New York Times* (February 25, 1971).
16. *Ibid.* (April 18, 1973).
17. C. P. Gilmore, "The Real Villain in Heart Disease," *New York Times Magazine* (March 25, 1973).
18. Kerr L. White, "Life and Death in Medicine," *Scientific American* (September 1973).

19. Robert S. Morrison, "Dying," *ibid.*
20. Alexander Leaf, "Getting Old," *ibid.*
21. *New York Times* (March 3, 1972).
22. John Knowles, "The Hospital," *Scientific American* (September 1973).

XIV   MISCELLANEOUS MEDIA

1. *New York Times* (December 5, 1972).
2. Roy Bongartz, "The Junk Mail Boom," *New York Times Magazine* (October 22, 1972).
3. *New York Times* (December 6, 1973).
4. *Ibid.* (January 2, 1973).

XV   THE WAY IT IS

1. Harriet Van Horne, "Views Coming Up," *Collier's* (October 13, 1945).
2. Cited in David W. Rintels, "But Will Marcus Welby, M.D., Always Make You Well?" *New York Times* (March 12, 1972).
3. *Village Voice* (April 20, 1971).
4. Donald Horton and R. Richard Wohl, "Mass Communications and Para-Social Interaction," from *America as a Mass Society*, edited by Philip Olson. New York, Free Press of Glencoe, 1963.
5. Erma Bombeck, "At Wits End," a newspaper column. Field Enterprises, 1974.
6. Professor Douglas Heath, from an address on November 13, 1974.
7. Rintels, *op cit.*
8. Cited in Edward J. Epstein, "Television Network News," *The New Yorker* (March 3, 1973).
9. *Ibid.*
10. *Ibid.*

XVI   THE ARTIFICIAL AMERICAN

1. Dr. Robert Jastrow, "Satellites May Make the Earth the Smallest of Worlds," *New York Times* (June 9, 1974).
2. *Ibid.*
3. *Ibid.*
4. Rona Cherry and Lawrence Cherry, "Depression," *New York Times Magazine* (November 25, 1973).

5. Slater, *op. cit.*
6. *New York Times* (July 25, 1973).
7. *Wall Street Journal* (October 24, 1974).
8. Ada Louise Huxtable, *New York Times* (March 21, 1971).
9. *New York Times* (October 13, 1972).
10. *Ibid.* (July 7, 1974).
11. J. Wreford Watson, "The Geography of Newness," *New York Times* (December 23, 1970).
12. Andrew H. Malcolm, "The New Highway 'Towns'—Nice Places to Go Through," *New York Times* (July 25, 1973).
13. Ehud Yonay, *New York Times* (November 26, 1972).
14. *New York Times* (May 4, 1972).
15. Archibald MacLeish, "Rediscovering the Simple Life," *McCall's* (April 1972).

About the Author

TRUMAN MOORE is a free-lance writer and photographer. His first book, *The Slaves We Rent*, is considered a classic study of migrant labor in America. His second book, *The Traveling Man*, was published in 1972.

In the last sixteen years the author has traveled widely in the United States doing book research and photographic assignments. *Nouveaumania* grew out of his impressions and observations of America during those years of almost constant travel.

Mr. Moore majored in journalism at the University of North Carolina and is currently doing graduate work in sociology at New York University. He lives in New York City with his wife and daughter.